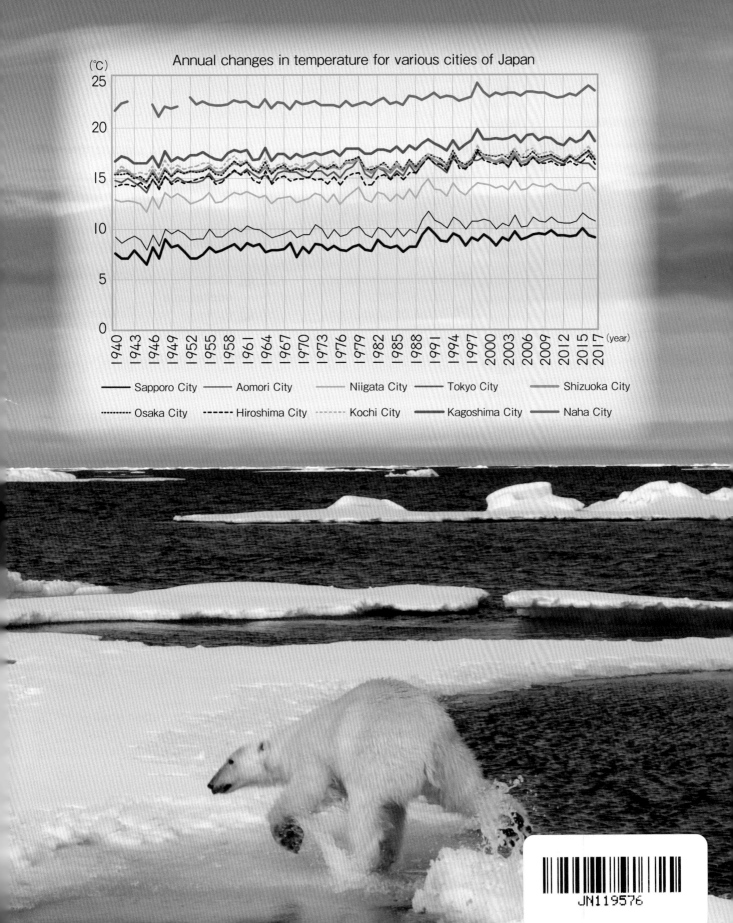

Annual changes in temperature for various cities of Japan

— Sapporo City — Aomori City — Niigata City — Tokyo City — Shizuoka City
······· Osaka City ----- Hiroshima City ----- Kochi City — Kagoshima City — Naha City

JN119576

Table of contents

3rd grade Decimal Numbers ▷ **10** Decimal Numbers

Let's explore how to represent decimal numbers ······················ 4
and their structure.

❶ How to represent ··········· 5 ❸ Addition and subtraction···· 16
decimal numbers of decimal numbers

❷ Structure of decimal ········ 12
numbers

3rd grade Addition and Subtraction ▷ **11** Math Expressions and Operations

Let's interpret math expressions using the rules of ···················· 22
operations.

❶ Math sentences and ········ 23 ❸ Using the rules of ············· 30
operations operations

❷ Rules of operations ········ 27 ❹ Rules of multiplication········ 32
and division

3rd grade Addition and Subtraction ▷ **12** Calculation with Whole Numbers

Let's summarize the ways of calculation with whole numbers.···· 36

1st grade Comparing Sizes ▷ **13** Area

Let's explore how to represent and how to find the size of area.··· 40

❶ Area ····························· 41 ❸ Units for large areas··········· 50

❷ Area of rectangles and ···· 46 ❹ Relationship between ········· 55
squares area units

Reflect・Connect ·· 58

14 Let's Think about How to Calculate

Let's think about judicious ways for how to calculate with ······· 60
decimal numbers.

❶ Decimal number × whole number····························· 61

❷ Decimal number ÷ whole number····························· 63

15 Multiplication and Division of Decimal Numbers

Let's think about how to multiply and divide decimal ··············· 65
numbers in vertical form.

❶ Calculation of decimal······ 65 ❸ Various divisions ·············· 72
number × whole number

❷ Calculation of decimal······ 69 ❹ Which operation shall ········ 75
number ÷ whole number we use?

How many times: Try Boccia. ·· 78

Utilizing rule of three on a 4-cell table ································· 80

Let's learn mathematics together.

Nanami

4th Grade. Volume 1

❶ Large Numbers ❹ Division by 1-digit Number ❼ Division by 2-digit Number
❷ Line Graph ❺ Angles ❽ Round Numbers
❸ The case: (2-digit)÷(1-digit) ❻ Perpendicular, Parallel, and ❾ Arrangement of Data
Quadrilaterals

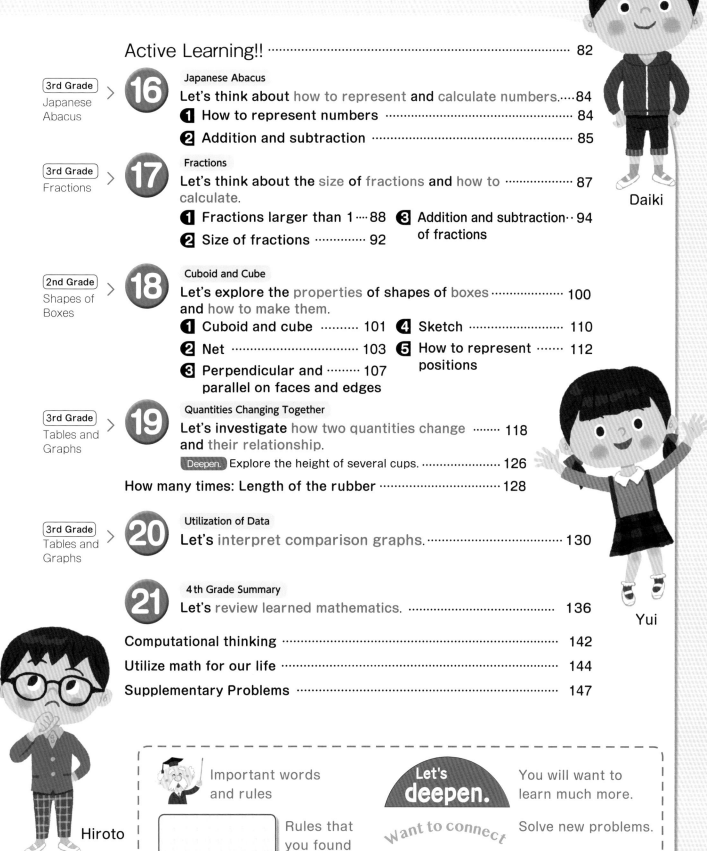

Active Learning!! 82

3rd Grade
Japanese Abacus

16 Japanese Abacus
Let's think about how to represent and calculate numbers.....84
❶ How to represent numbers 84
❷ Addition and subtraction 85

3rd Grade
Fractions

17 Fractions
Let's think about the size of fractions and how to 87
calculate.
❶ Fractions larger than 1 ....88 ❸ Addition and subtraction·· 94
❷ Size of fractions 92   of fractions

2nd Grade
Shapes of Boxes

18 Cuboid and Cube
Let's explore the properties of shapes of boxes 100
and how to make them.
❶ Cuboid and cube .... 101 ❹ Sketch 110
❷ Net .... 103 ❺ How to represent 112
❸ Perpendicular and 107   positions
   parallel on faces and edges

3rd Grade
Tables and Graphs

19 Quantities Changing Together
Let's investigate how two quantities change 118
and their relationship.
Deepen. Explore the height of several cups. 126

How many times: Length of the rubber 128

3rd Grade
Tables and Graphs

20 Utilization of Data
Let's interpret comparison graphs. 130

21 4th Grade Summary
Let's review learned mathematics. 136

Computational thinking 142

Utilize math for our life 144

Supplementary Problems 147

Daiki

Yui

Hiroto

Important words and rules

Rules that you found

Let's **deepen.**

Want to connect

You will want to learn much more.

Solve new problems.

Find the ?

Can you say how many liters?

Let's play the 1 L game.

How to play

① Let's try to pour 1 L of water in the kettle.

② The person who pours closest to 1 L wins.

If I represent the amount of water by liters...

Hiroto

1 L

1 L
remaining

Mine is closer to 1 L than Hiroto's, but if I represent by liters...

Yui

1 L

1 L
remaining

I can't read the remaining amount of water of Yui.

What should we do?

Problem By how many liters can the amount of water be represented?

10 Decimal Numbers

Let's explore how to represent decimal numbers and their structure.

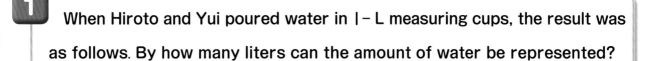

1 How to represent decimal numbers

Want to know Amount of water that is legs than 0.1 L

1 When Hiroto and Yui poured water in 1 – L measuring cups, the result was as follows. By how many liters can the amount of water be represented?

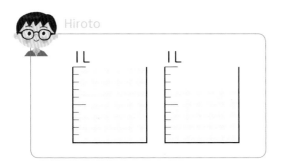

Hiroto

1 L   1 L

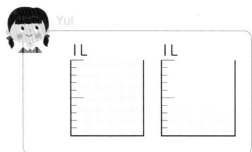

Yui

1 L   1 L

① As for Hiroto's amount of water, it is 1 L and ☐ sets of 0.1 L. Therefore, ☐ L.

② As for Yui's amount of water, can you say how many liters?

Way to see and think

What are you considering as one unit for the remaining amount of water?

1 L

0.1 L

Daiki

I can't measure with a 0.1 L scale.

When the remaining amount of water couldn't fit in 1 L, we divided 1 L into 10 equal parts.

Nanami

Purpose When the remaining amount of water is less than 0.1 L, how should we represent it?

③ Let's try to measure the amount of water that is less than 0.1 L by making a smaller unit scale. We can divide 0.1 L into 10 equal parts.

Way to see and think

The same as when 0.1 L was made, divide into 10 equal parts.

④ Let's represent Yui's amount of water.

☐ . ☐ ☐ L

Number of 1 – L measuring cup | Number of 0.1 – L measuring cup | Number of small unit

⑤ How many liters is the amount of 1 small unit scale?

0.1 L

☐ . ☐ ☐ L

Number of 1 – L measuring cup | Number of 0.1 – L measuring cup | Number of small unit

The amount that is obtained by dividing 0.1 L into 10 equal parts is written as 0.01 L and is read as **one hundredth liter** or **zero point zero one liter**.

1 dL = 0.1 L

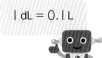

Yui's amount of water is 1.36 L and is read as "one point three six liters."

1 of 1L	is 1L
3 of 0.1L	is 0.3L
6 of 0.01L	is 0.06L
Altogether:	1.36L

Summary

An amount that is less than 0.1 L can be represented by how many 0.01 L, which is found after dividing 0.1 L into 10 equal parts.

Length that is smaller than 0.1 m

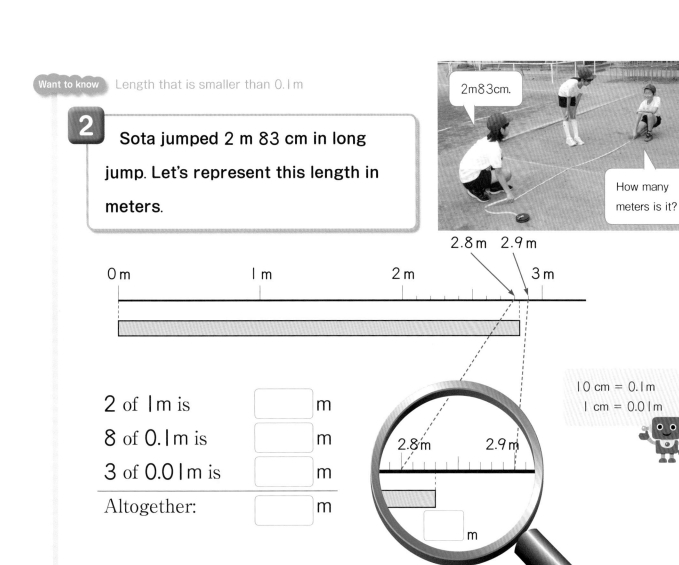

2 Sota jumped 2 m 83 cm in long jump. Let's represent this length in meters.

2m83cm.

How many meters is it?

2.8 m   2.9 m

0 m   1 m   2 m   3 m

2 of 1m is ☐ m

8 of 0.1m is ☐ m

3 of 0.01m is ☐ m

Altogether: ☐ m

2.8 m   2.9 m

☐ m

10 cm = 0.1 m
1 cm = 0.01 m

1 How many liters are the following amounts of water?

① 1L   0.1L 0.1L 0.1L

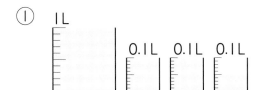

② 1L   0.1L 0.1L 0.1L

2 Let's read the following numbers marked by ↑.

2.9   3   3.1   3.2(m)

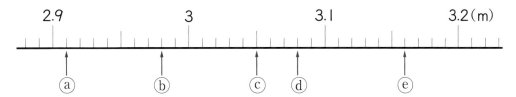

ⓐ   ⓑ   ⓒ ⓓ   ⓔ

Summary notebook

Let's summarize what you learned in the day.

Become
a writing
master.

Write today's date.

October 15

Write the problem.

When Hiroto and Yui poured water in 1−L measuring cups, the result was as follows. The amount of water can be represented by how many liters?

Let's use diagrams to organize the problem.

Hiroto
1 L and
7 sets of 0.1 L

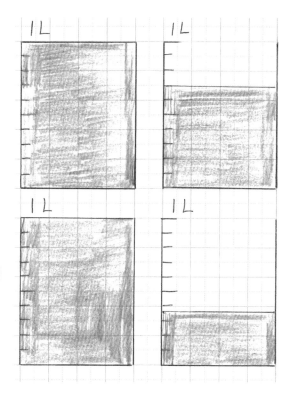

Yui
1 L,
3 sets of 0.1 L, and
a little more

Let's learn based on the purpose.

Purpose   When the remaining amount of water is less than 0.1 L, how should we represent it?

〈My own idea〉

Place a small unit scale by dividing 0.1 L into 10 equal parts.

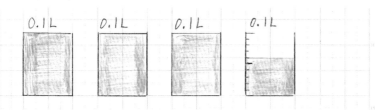

1 of   1 L   is   1 L
3 of   0.1 L   is 0.3 L
6 of   0.01 L is 0.06 L
_____
         Altogether: 1.36 L

Summary

An amount that is less than 0.1 L can be represented by how many 0.01 L, which is found after dividing 0.1 L into 10 equal parts.

〈Reflection〉

When there is a remaining amount less than 0.01 L, can we also divide into 10 equal parts?

〈Marin's notebook〉

		1	.			L
1 of	1 L	is	1			L
3 of	0.1 L	is	0.	3		L
6 of	0.01 L	is	0.	0	6	L
	Altogether:		1.	3	6	L

Let's organize and write your own ideas.

Let's write today's summary surrounded by colored lines.

As for reflection, the following must be written:
· understood things,
· noticed things,
· what you are able to do,
· not understood things,
· what you want to do more.

Marin summarized by place value.

3 **Let's represent the following amount of water in liters**.

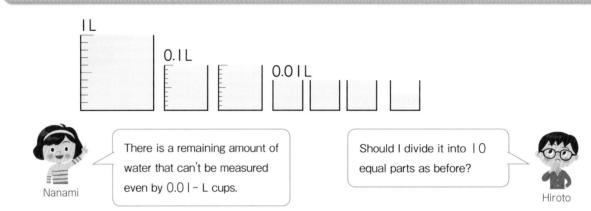

Nanami: There is a remaining amount of water that can't be measured even by 0.0 l – L cups.

Hiroto: Should I divide it into 10 equal parts as before?

🔲 **Purpose** When the remaining amount of water is less than 0.0 l L, how should we represent it?

Want to improve

① Let's try to measure the amount of water that is less than 0.01 L by making a smaller unit scale. We can divide 0.01 L into 10 equal parts.

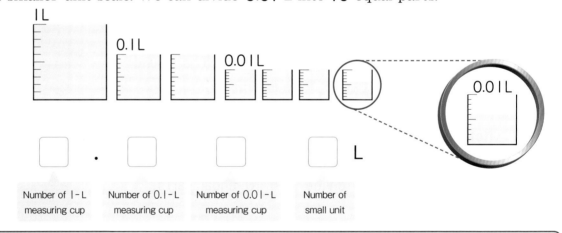

Number of 1‑L measuring cup	Number of 0.1‑L measuring cup	Number of 0.01‑L measuring cup	Number of small unit

The amount that is obtained by dividing 0.01 L into 10 equal parts is written as 0.001L and is read as **one thousandth liter** or **zero point zero zero one liter**.

🔲 **Summary**

An amount that is less than 0.0 l L can be represented by how many 0.00 l L, which is found after dividing 0.0 l L into 10 equal parts.

3 The length of the bridge connecting Miyakojima and Ikemajima in Okinawa Prefecture is 1425 m. Let's represent this length in kilometers.

1 of 1km is	1 km
4 of 0.1km is	☐ km
2 of 0.01km is	☐ km
5 of 0.001km is	☐ km
Altogether:	☐ km

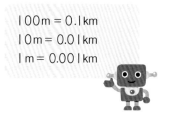

100m = 0.1km
10m = 0.01km
1m = 0.001km

4 Let's represent 1kg 264g in kilograms.

1 of 1 kg is	1 kg
2 of 0.1 kg is	☐ kg
6 of 0.01 kg is	☐ kg
4 of 0.001 kg is	☐ kg
Altogether:	☐ kg

100 g = 0.1kg
10 g = 0.01kg
1 g = 0.001kg

 5 Let's represent the following measurements by using the unit shown in ( ).

① 1435 mm (m)　　② 42195 m (km)　　③ 875 g (kg)

Want to explore

1 Let's explore the relationship between 1, 0.1, 0.01, and 0.001.

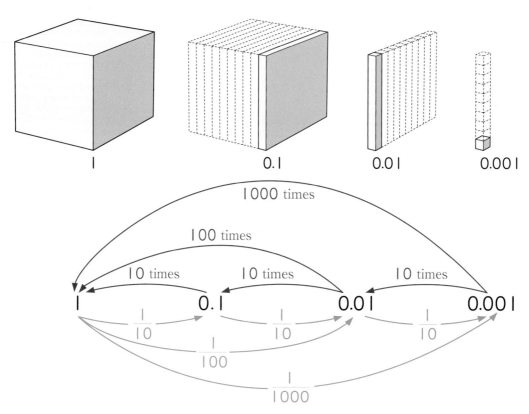

① How many times of 0.1 is 1? Also, what fraction of 1 is 0.1?

② Let's explore the relationship between 0.01 and 1.

③ Let's explore the relationship between 0.001 and 1.

④ Let's explore other relationships such as the one between 0.1 and 0.01.

Want to think

How many sets of 1, 0.1, 0.01, and 0.001 are altogether in the number 2.386?

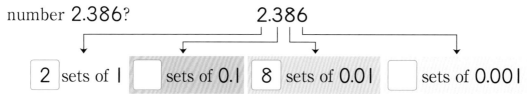

The places immediately to the right of the decimal point are as follows:

Tenths place $\left(\dfrac{1}{10} \text{ place} \right)$,

Hundredths place $\left(\dfrac{1}{100} \text{ place} \right)$,

Thousandths place $\left(\dfrac{1}{1000} \text{ place} \right)$.

Decimal numbers, just like whole numbers, are represented by ten times or $\dfrac{1}{10}$ of a place.

$2 \cdot 386$

...ones place
...decimal point
...tenths place
...hundredths place
...thousandths place

Want to try

 2 Let's write the following numbers.

Way to see and think

How many are there for each place value?

① The number that is the sum of 3 sets of 1, 4 sets of 0.1, 9 sets of 0.01, and 5 sets of 0.001.

② The number that is the sum of 7 sets of 1, 3 sets of 0.1, and 5 sets of 0.001.

Want to explore

 3 Let's examine how many sets of 0.001 are gathered in 3.254.

Way to see and think

1 is a number that gathers 10 sets of 0.1 and 0.1 is a number that gathers 10 sets of 0.01.

3 gathers		sets of 0.001
0.2 gathers		sets of 0.001
0.05 gathers		sets of 0.001
0.004 gathers		sets of 0.001
3.254 gathers		sets of 0.001

Want to confirm

 4 How many sets of 0.01 are gathered in the following numbers?

① 1.27   ② 0.58   ③ 2

13

2 Let's compare the size of the following numbers.

Ⓐ 0.43   Ⓑ 0.425   Ⓒ 0.432

Daiki

When it's a whole number, we compare from the highest place value.

What should be a good way to compare?

Nanami

Ⓨ Purpose **How should we compare the size of decimal numbers?**

Want to compare

① Let's write each of the numbers on the table and compare the sizes.

Ones place	$\frac{1}{10}$ place	$\frac{1}{100}$ place	$\frac{1}{1000}$ place
Ⓐ			
Ⓑ			
Ⓒ			

Way to see and think

Think about aligning the place value.

② Let's compare the size by marking the position with a ↑ in the following number line.

Way to see and think

How much is represented by one unit scale?

0.4   0.5

Ⓠ Summary

The size of decimal numbers can be compared by aligning the decimal point or representing the decimal numbers in the number line.

Want to try

 Let's put the following numbers in descending order.

0.5   5   0.005   0   0.05

3 What number is 10 times of 0.58? Also, what number is $\frac{1}{10}$ of 0.58?

We can calculate 10 times and $\frac{1}{10}$ of a whole number.

When it's a whole number, getting 10 times increases one 0 and getting $\frac{1}{10}$ decreases one 0.

Yui

Hiroto

Purpose What kind of number do 10 times and $\frac{1}{10}$ of a decimal number become?

① Let's write 10 times and $\frac{1}{10}$ of the number in the table.

Let's discuss what kind of number it becomes.

Ones place	$\frac{1}{10}$ place	$\frac{1}{100}$ place	$\frac{1}{1000}$ place	
0 .	5	8		) 10 times
				) $\frac{1}{10}$

Summary

Like whole numbers, a decimal number in numerals, when multiplied by 10, becomes a number in the next higher place. And $\frac{1}{10}$ of a decimal number in numerals becomes a number in the next lower place.

 6 Let's write the number that is 10 times, 100 times, 1000 times, and $\frac{1}{10}$ of 3.92.

Thousands place	Hundreds place	Tens place	Ones place	$\frac{1}{10}$ place	$\frac{1}{100}$ palce	$\frac{1}{1000}$ place	
							← 1000 times
							← 100 times
							↖ 10 times
		3 .	9	2			
							) $\frac{1}{10}$

 7 Let's find the number that is 10 times and $\frac{1}{10}$ of the following numbers.

① 0.74   ② 1.58   ③ 26.95

Want to solve   Addition of decimal numbers

Activity

1 A tank contains 2.25 L of water. When 1.34 L of water is poured into it, how many liters are there in total?

① Let's write a math expression.

I did the addition of decimal numbers until the tenths place.

Daiki

Can I calculate in the same way even until the hundredths place?

Nanami

▼ **Purpose**   Can we calculate the addition even until the hundredths place?

Want to explain

② Let's explain the calculation method of the two children.

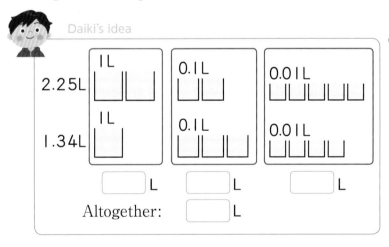

Daiki's idea

2.25L

1.34L

[ ] L   [ ] L   [ ] L

Altogether: [ ] L

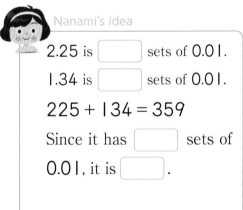

Nanami's idea

2.25 is [ ] sets of 0.01.

1.34 is [ ] sets of 0.01.

225 + 134 = 359

Since it has [ ] sets of

0.01, it is [ ].

Addition algorithm to calculate 2.25 + 1.34 in vertical form

$$
\begin{array}{r}
2.2\,5 \\
+\ 1.3\,4 \\
\hline
\end{array}
$$

→

$$
\begin{array}{r}
2.2\,5 \\
+\ 1.3\,4 \\
\hline
3\ 5\ 9
\end{array}
$$

→

$$
\begin{array}{r}
2.2\,5 \\
+\ 1.3\,4 \\
\hline
3.5\,9
\end{array}
$$

Align the digits according to their places.

Calculate the numbers in each place in the same way as whole numbers.

Place the decimal point of the sum in the same position as the above decimal point.

Summary

In the addition of decimal numbers, the same as for the case of whole numbers, calculate by aligning the digits according to their places.

Want to confirm

 1 Let's solve the following calculations in vertical form.

① 2.16 + 0.73

```
    2 . 1  6
 +  0 . 7  3
```

② 5.74 + 2.63

What should I do when 0 is in the math expression or the answer?

Want to think

 2 Let's think about how to solve the following calculations in vertical form.

① 9.23 + 0.47

② 5.04 + 2.96

Hiroto

Want to explain

3 Is the calculation in vertical form on the right correct? If it is not, let's explain how to calculate correctly in vertical form.

```
   4.05
 +  3.1
   4.36
```

 Way to see and think

Is the digits of the numbers properly aligned?

Want to confirm

 4 Let's solve the following calculations in vertical form.

① 6.27 + 3.51   ② 8.46 + 0.32   ③ 1.54 + 2.38

④ 4.72 + 3.49   ⑤ 9.62 + 0.18   ⑥ 3.21 + 2.5

2 Tank Ⓐ contains 3.46 L of water. Tank Ⓑ contains 2.14 L of water.

How many more liters does Tank Ⓐ contain than Tank Ⓑ?

① Let's write a math expression.

Ⓨ Purpose Can the subtraction of decimal numbers be calculated in the same way as the addition?

Want to explain

② Let's explain Daiki's idea. Also, let's find the answer by using this idea.

Daiki's idea

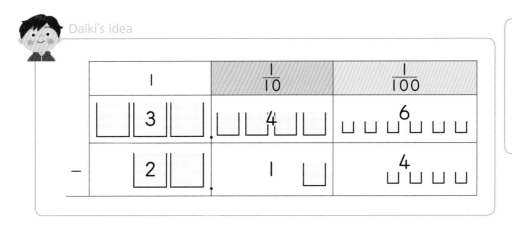

We can also calculate by considering how many sets of 0.01 are gathered in each number.

Nanami

Subtraction algorithm to calculate 3.46 − 2.14 in vertical form

```
  3.4 6
− 2.1 4
```
Align the digits according to their places.

➡

```
  3.4 6
− 2.1 4
  1.3 2
```
Calculate the numbers in each place in the same way as whole numbers.

➡

```
  3.4 6
− 2.1 4
  1.3 2
```
Place the decimal point of the difference in the same position as the above decimal point.

🌼 Summary

In the subtraction of decimal numbers, the same as for the case of whole numbers, calculate by aligning the digits according to their places.

Want to confirm

 5 ▶ Let's solve the following calculations in vertical form.

① 5.78 − 3.44   ② 1.54 − 0.23   ③ 8.37 − 2.09

Want to think

 6 ▶ Let's think about how to solve the following calculations in vertical form.

Way to see and think

① 2.32 − 1.82

```
    2 . 3  2
−   1 . 8  2
```

② 6.7 − 3.91

When there is borrowing, the same as with whole numbers, you should borrow from a higher place.

What should I do when 0 is in the math expression or the answer?

③ 6 − 0.52

④ 5.03 − 4.25

Yui

Want to confirm

7 ▶ Let's solve the following calculations in vertical form.

① 0.54 − 0.34   ② 1.96 − 0.56   ③ 7.28 − 2.4
④ 4 − 1.26   ⑤ 3.4 − 1.84   ⑥ 7.08 − 6.28

Want to try

 8 ▶ 85 cm were cut out from a 2.15 m tape. How many meters is the remaining length of the tape?

What you can do now

☐ **Understanding how to represent decimal numbers.**

1 How many liters are the following amounts of water?

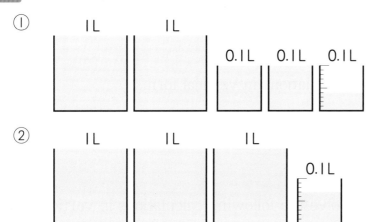

① | 1 L | 1 L | 0.1 L | 0.1 L | 0.1 L

② | 1 L | 1 L | 1 L | 0.1 L

☐ **Understanding the structure of decimal numbers.**

2 Let's fill in the following ☐ with numbers.

① 8.61 is the number that is the sum of 8 sets of ☐, 6 sets of ☐, and 1 set of ☐.

② ☐ is 10 times of 0.46, and ☐ is $\frac{1}{10}$ of 0.46.

③ 1 g is ☐ kg.

④ 1 L is ☐ mL.

⑤ 1 m is ☐ km.

☐ **Understanding the size of decimal numbers.**

3 Let's fill in the ☐ with appropriate inequality signs.

① 0.21 ☐ 0.189   ② 2.395 ☐ 2.5

☐ **Can add and subtract decimal numbers.**

4 Let's solve the following calculations in vertical form.

① 5.32 + 1.45   ② 4.18 + 0.32   ③ 3.64 + 2.4

④ 10.8 + 3.45   ⑤ 9.26 − 4.12   ⑥ 7.05 − 4.6

⑦ 3 − 1.29   ⑧ 0.87 − 0.17

Supplementary Problems
p.148

Usefulness and efficiency of learning

1 Let's represent the following measurements by using the unit shown in ( ).

① 8695 g (kg)　　　② 320 mL (L)

③ 3.67 km (m)

Understanding how to represent decimal numbers.

2 Let's answer the following problems.

① Let's write the number that is the sum of 6 sets of 1, 4 sets of 0.1, 9 sets of 0.01, and 3 sets of 0.001.

② Let's write each of the numbers that are 10 times, 100 times, 1000 times, and $\frac{1}{10}$ of 2.79 and 18.83.

Understanding the structure of decimal numbers.

Understanding the structure of decimal numbers.

3 Let's put the following numbers in descending order.

3.15　　0.98　　1.05　　2.8　　6.12

Understanding the size of decimal numbers.

4 At my first attempt in long jump, I jumped 2.81 m. In the second attempt, my jump record was 25 cm longer than the first jump. How many meters did I jump at the second attempt?

Can add and subtract decimal numbers.

5 There are 4 kg of wheat. Yesterday I used 0.5 kg, and today I used 0.35 kg. How many kilograms are remaining?

Can add and subtract decimal numbers.

6 If three ribbons with length 2.17 m, 3.62 m, and 2.45 m are connected, how many meters does it become in total?

Can add and subtract decimal numbers.

If we represent it in a math sentence...?

Problem What kind of math sentence will make the calculation easier?

Math Expressions and Operations

11 Let's interpret math expressions using the rules of operations.

Want to solve Represented in one math sentence

1 Nanami is going shopping with a 500-yen coin. She will buy a notebook for 120 yen at a stationery store and batteries for 360 yen at an electrical store. How many yen will be remaining?

① Let's represent the ideas of the following children in math sentences.

 Nanami's idea

First, I find the remaining amount of money after buying the notebook. Then, I subtract the cost of the batteries from the remaining amount.

$$500 - \boxed{} = \boxed{}$$
$$\boxed{} - 360 = \boxed{}$$

 Hiroto's idea

First, I think how much is the notebook and the batteries altogether. Then, I subtract the total cost from the 500 yen.

$$120 + 360 = \boxed{}$$
$$500 - \boxed{} = \boxed{}$$

Want to discuss

② Let's try to discuss how the math sentences in ① can be represented in one math sentence.

 Since Nanami buys one by one in order...

Daiki

Hiroto is thinking about the total cost of what he buys.

Yui

Purpose How should we represent two math sentences in one math sentence?

③ Let's represent Nanami's idea in one math sentence.

$$500 - \boxed{} - \boxed{} = \boxed{}$$

④ Let's represent Hiroto's idea in one math sentence.

$$500 - (\boxed{}) = \boxed{}$$

Money brought   Total cost   Money left

⚘ Summary

As in the total cost, using the symbol ( ) can be considered together, so that two math sentences can be represented in one math sentence.

In math sentences with ( ), the calculation inside ( ) is done first.

$$500 - (120 + 360) = 500 - 480$$
$$\underset{(2)}{\underline{\qquad \underset{(1)}{\underline{\qquad}} \qquad}} = 20$$

1 A snack priced at 350 yen is sold with a 50 yen discount. When one snack is bought with 1000 yen, how many yen is the change? Let's use ( ) and find the answer by representing this question with a math sentence.

$$\boxed{} - (\boxed{}) = \boxed{}$$

Amount paid   Cost of a snack   Change

2 Let's make problems for the following math expressions.

① 400 − (50 + 300)   ② 600 − (150 − 110)

2

Yua bought a racket for 900 yen and two

shuttlecocks for 100 yen each.

What is the total cost?

① Let's write one math expression to find the total cost.

② Let's think about the order of operations.

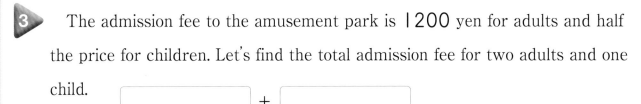

$$900 \ + \ 100 \times 2$$

Cost of a racket   Cost of shuttlecocks

Nanami

Since the cost of the shuttlecocks are considered together...

Want to try

3 The admission fee to the amusement park is 1200 yen for adults and half the price for children. Let's find the total admission fee for two adults and one child.

[   ] + [   ]

Admission fee for 2 adults   Admission fee for 1 child

In a math expression that includes addition, subtraction, multiplication, and division, the multiplication and division are done first even without ( ).

Want to confirm

4 Let's solve the following calculations.

① $800 - 200 \times 3$ ② $24 - 12 \div 4$ ③ $8 \times 5 + 20 \div 5$

3   Let's solve the following calculations, but be careful about the order of operations.

① $12 + 15 \div (5 - 2)$ ② $12 + (15 \div 5 - 2)$

① $12 + 15 \div \underset{(1)}{(5 - 2)} = 12 + \underset{(2)}{15 \div 3}$
$= 12 + 5 \,^{(3)}$
$= \boxed{}$

$12 + 15 \div (5 - 2)$
(1)
(2)
(3)

② $12 + \underset{(1)}{(15 \div 5} - 2) = 12 + \underset{(2)}{(3 - 2)}$
$= 12 + 1 \,^{(3)}$
$= \boxed{}$

$12 + (15 \div 5 - 2)$
(1)
(2)
(3)

If you write the math expressions in correct order using equal signs as shown above, the solution can be easier to understand.

Order of operations

(1) A math expression is basically solved from the left.

(2) If ( ) is included, the calculation inside ( ) is done first.

(3) In a math expression that includes $+$, $-$, $\times$, and $\div$, do the multiplication and division first.

---

5   Let's solve the following calculations.

① $12 \div 2 \times 3$ ② $12 \div (2 \times 3)$

③ $(5 + 4) \times (6 - 2)$ ④ $5 + 4 \times (6 - 2)$

⑤ $90 - 50 \div (4 + 6)$ ⑥ $(90 - 50) \div 4 + 6$

Want to explore

 Calculations Ⓐ and Ⓑ were each calculated as shown in Ⓒ and Ⓓ.

What kind of rule was used to solve each calculation?

Ⓐ $5 + 397$ → Ⓒ $397 + 5$

Ⓑ $389 + 234 + 266$ → Ⓓ $389 + (234 + 266)$

In addition, there was a rule in which you could change the order of operation and calculate.

Daiki

If Ⓑ is changed to Ⓓ, it becomes easier.

Yui

In addition, there are rules as shown below:

(1)   The sum does not change even when the order of the augend and addend is changed.

■ + ▲ = ▲ + ■

(2)   When adding three numbers, the sum does not change even if you change their order.

(■ + ▲) + ● = ■ + (▲ + ●)

 ■, ▲, and ● each are replaced with the same number.

Want to confirm

 Let's write the following numbers that apply to ■, ▲, and ● in the math sentences shown in (1) and (2), and confirm that the rules of operation hold.

① 12 for ■, 21 for ▲, and 9 for ●

② 3.8 for ■, 2.3 for ▲, and 2.7 for ●

 Way to see and think

Do the rules of operation hold for decimal numbers as well as for whole numbers?

2 Calculations Ⓐ and Ⓑ were each calculated as shown in Ⓒ and Ⓓ. What kind of rule was used to solve each calculation?

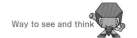

Way to see and think

What kind of rules did the multiplication have?

Ⓐ 55 × 248 → Ⓒ 248 × 55

Ⓑ 18 × 25 × 4 → Ⓓ 18 × (25 × 4)

In multiplication, there are rules as shown below:

(1) The product does not change even when the order of the multiplicand and multiplier is changed.

$$\blacksquare \times \blacktriangle = \blacktriangle \times \blacksquare$$

(2) When multiplying three numbers, the product does not change even if you change their order.

$$(\blacksquare \times \blacktriangle) \times \bullet = \blacksquare \times (\blacktriangle \times \bullet)$$

3 Let's write 3 for ■ , 5 for ▲ , and 4 for ● in the math sentences shown in (1) and (2) to calculate them, and confirm that the rules of operation hold.

Activity

2 There are two sheets of stickers as shown on the right. Let's think the following about these stickers.

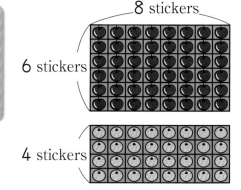

8 stickers

6 stickers

4 stickers

① If we combine all the stickers from the two sheets, how many stickers are there altogether? Let's compare the ideas of the following children.

Nanami's idea

$$6 \times \boxed{} + 4 \times \boxed{} = 48 + \boxed{}$$
$$= \boxed{}$$

Hiroto's idea

$$(6 + \boxed{}) \times 8 = \boxed{} \times 8$$
$$= \boxed{}$$

② How many stickers is the difference between the number of stickers on the two sheets? Let's compare the ideas of the following children.

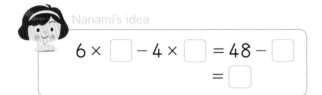

Nanami's idea

$$6 \times \boxed{} - 4 \times \boxed{} = 48 - \boxed{}$$
$$= \boxed{}$$

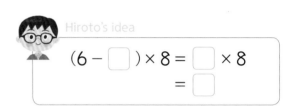

Hiroto's idea

$$(6 - \boxed{}) \times 8 = \boxed{} \times 8$$
$$= \boxed{}$$

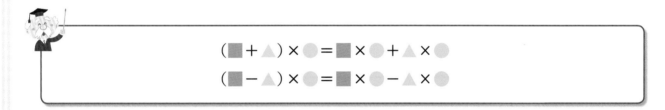

$$(\blacksquare + \triangle) \times \bullet = \blacksquare \times \bullet + \triangle \times \bullet$$
$$(\blacksquare - \triangle) \times \bullet = \blacksquare \times \bullet - \triangle \times \bullet$$

4 Let's write 5 for $\blacksquare$, 4 for $\triangle$, and 3 for $\bullet$ in the math sentences shown above, and confirm that the rules of operation hold.

5 Let's solve the following calculations.

① $(4 + 16) \times 3$

② $5 \times (14 - 9)$

③ $25 \times 4 + 15 \times 4$

④ $30 \times 7 - 28 \times 7$

1

I went to buy goldfish that cost 200 yen each. Since the store gave a 20 yen discount for each goldfish, I bought 6 goldfish. How much was the total cost?

① To find the cost, Daiki and Yui thought of math expressions shown below. Let's explain each idea.

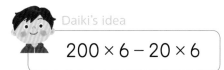

Daiki's idea

$$200 \times 6 - 20 \times 6$$

Yui's idea

$$(200 - 20) \times 6$$

② Let's explain that their answers are the same by using the rules of operations.

If you observe closely the ( ) and order of operations, you can understand the way of thinking when the math expression was made. Also, if you use the rules of operations, you can explain the reason why the answers of the two math expressions are the same without finding the answer.

To find the number of marbles that are lined up as shown on the right, the math expressions ① and ② were made. Let's discuss how both math expressions were considered.

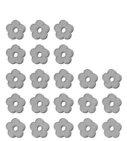

① $2 \times 3 + 3 \times 5$ ② $5 \times 5 - 2 \times 2$

2   Let's use the rules of operations and improve the calculation.

① $76 + 47 + 53 = 76 + (\boxed{} + \boxed{})$

$= 76 + \boxed{}$

$= \boxed{}$

Way to see and think

Thinking if we can have an easier number to calculate.

② $18 \times 5 = (\boxed{} \times 2) \times 5$

$= \boxed{} \times (2 \times 5)$

$= \boxed{} \times 10$

$= \boxed{}$

Way to see and think

If 10 is made, the calculation is easier.

③ $198 \times 3 = (\boxed{} - 2) \times 3$

$= \boxed{} \times 3 - 2 \times 3$

$= \boxed{} - 6$

$= \boxed{}$

Way to see and think

Calculating by changing 198 to an easier representation.

 3   Let's solve the following calculations by using the rules of operations.

① $23 + 49 + 77$ ② 46×5 ③ 99×7

 4   Let's solve the following calculations and compare the math expressions and their products.

101×36 101×79

① Let's look at the two math expressions and their products, and discuss what you noticed.

② Let's explain what you noticed by using the rules of operations.

$101 \times 36 = (100 + 1) \times 36$, therefore...

Yui

Want to find Rules of multiplication

1 What is the total cost if I buy ▢ gums that cost 40 yen each?

Let's explore what kind of rules there are between the number ▢

and the answer, when various numbers are placed in ▢ .

$$40 \times ▢$$

$40 \times 3 = 120$ $40 \times 12 = 480$

$40 \times 15 = 600$ $40 \times 6 = 240$

$40 \times 6 = 240$ $40 \times 12 = 480$

$\downarrow_{\times ▢} \quad \downarrow_{\times ▢}$ $\downarrow_{\div ▢} \quad \downarrow_{\div ▢}$

$40 \times 12 = 480$ $40 \times 6 = 240$

In multiplication, if the multiplier is done ▢ times, then the product also becomes ▢ times.

Also, if the multiplier is divided by ▢ , then the product also becomes a number divided by ▢ .

Want to try

1 Let's find various rules about the multiplicand and multiplier, and about the multiplicand and product.

① $40 \times 6 = 240$

$\downarrow_{\times ▢} \quad \downarrow_{\div ▢}$

$80 \times 3 = 240$

② $80 \times 3 = 240$

$\downarrow_{\div ▢} \quad \downarrow_{\times ▢}$

$40 \times 6 = 240$

③ $40 \times 6 = 240$

$\downarrow_{\times ▢} \quad \downarrow_{\times ▢}$

$80 \times 6 = 480$

④ $80 \times 6 = 480$

$\downarrow_{\div ▢} \quad \downarrow_{\div ▢}$

$40 \times 6 = 240$

2 Let's place various numbers in ☐ of 24 ÷ ☐ , and explore the rules between the divisor and the quotient.

24 ÷ 4 = 6

×☐ ÷☐

24 ÷ 8 = 3

24 ÷ 6 = 4

÷☐ ×☐

24 ÷ 3 = 8

> In division, if the divisor is done ☐ times, then the quotient becomes a number divided by ☐.
>
> Also, if the divisor is divided by ☐, then the quotient becomes ☐ times.

Want to try

 3 Let's find the rules between the divident and the quotient.

12 ÷ 3 = 4

×☐ ×☐

24 ÷ 3 = 8

27 ÷ 3 = 9

÷☐ ÷☐

9 ÷ 3 = 3

> In division, if the dividend is done ☐ times, then the quotient also becomes ☐ times.
>
> Also, if the dividend is divided by ☐, then the quotient also becomes a number divided by ☐.

Want to confirm

 4 Let's try to confirm the rules found above with other multiplications and divisions.

What you can do now

Understanding the order of operations.

1 Let's solve the following calculations.

① $500 - (80 + 250)$ ② $650 - (430 - 60)$

③ $(40 + 50) \times 7$ ④ $6 \times (18 - 3)$

⑤ $120 \div (12 - 4)$ ⑥ $(37 + 18) \div 5$

⑦ $(11 - 4) \times (8 + 7)$ ⑧ $(14 + 22) \div (9 - 5)$

⑨ $18 \times 8 \div 4$ ⑩ $18 \times (8 \div 4)$

⑪ $28 - 3 \times (13 - 8)$ ⑫ $(32 - 18) + 4 \times 5$

⑬ $8 + 12 \times 3$ ⑭ $40 - 12 \div (6 \div 2)$

⑮ $40 \times 8 - 5 \times 24$ ⑯ $36 + 6 \times 8 \div 12$

Understanding the rules of operations.

2 Let's represent the following problems in one math sentence and find the answer.

① There were 60 sheets of drawing paper. I used 15 sheets yesterday, and 20 sheets today. How many sheets are remaining?

$60 - (\boxed{} + \boxed{}) = \boxed{}$

② There were 5 dozen of pencils. From those, I used 40 pencils. How many pencils are remaining?

$\boxed{} \times 5 - \boxed{} = \boxed{}$

③ There are 100 sheets of colored paper. 4 sheets for each person are distributed to a total of 18 people. How many sheets are remaining?

$\boxed{} - 4 \times \boxed{} = \boxed{}$

④ One set includes one pencil that costs 20 yen each and one eraser that costs 50 yen each. In total, 15 sets were made. What is the total cost of the sets?

$(\boxed{} + \boxed{}) \times 15 = \boxed{}$

Supplementary Problems
p.149

Usefulness and efficiency of learning

1 The following math expressions ① and ② were created to find the number of ◯ in the diagram shown on the right. Let's explain how the math expressions were considered by using each drawing as a source.

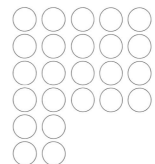

① $6 \times 2 + 4 \times 3$　　② $6 \times 5 - 2 \times 3$

2 Let's fill in each ☐ with a number.

① $25 \times 98 = 25 \times (\boxed{} - 2)$
$ = 25 \times \boxed{} - 25 \times 2$
$ = \boxed{} - \boxed{}$
$ = \boxed{}$

② $25 \times 24 = 25 \times \boxed{} \times 6$
$ = \boxed{} \times 6$
$ = \boxed{}$

③ $105 \times 6 = (\boxed{} + 5) \times 6$
$ = \boxed{} \times 6 + 5 \times \boxed{}$
$ = \boxed{} + \boxed{}$
$ = \boxed{}$

④ $99 \times 9 = (\boxed{} - 1) \times 9$
$ = \boxed{} \times 9 - 1 \times 9$
$ = \boxed{} - \boxed{}$
$ = \boxed{}$

3 Let's complete the math sentences by placing $+$, $-$, $\times$, or $\div$ in each ☐. When ( ) are needed, you may also include them.

① $4 \boxed{} 4 \boxed{} 4 \boxed{} 4 = 0$　　② $4 \boxed{} 4 \boxed{} 4 \boxed{} 4 = 1$

③ $4 \boxed{} 4 \boxed{} 4 \boxed{} 4 = 2$　　④ $4 \boxed{} 4 \boxed{} 4 \boxed{} 4 = 3$

Calculation with Whole Numbers

12 Let's summarize the ways of calculation with whole numbers.

Want to solve

1 In 2016, there were 556234 boys and 531768 girls in 4th grade. What was the total number of children in 4th grade?

① Let's write a math expression.

② Can you say approximately how many ten thousand children?

You should think of it by the nearest ten thousand round number.

Hiroto

③ Let's think about how to calculate.

Daiki

There are more digits than before.

Can I calculate in vertical form?

Nanami

Purpose How should we calculate an addition with many digits?

④ Let's explain Yui's idea.

Can you think of it in the same way as if adding 3-digit numbers?

Also, let's actually calculate.

Yui's idea

Until now, addition has been done in vertical form. Even if the number of digits increases, it can be calculated by each place value.

	5	5	6	2	3	4
+	5	3	1	7	6	8

Want to try

From exercise **1**, which are more, boys or girls? By how many?

① Let's write a math expression.

−				

② Let's calculate in vertical form.

Summary

In addition and subtraction, even if the number of digits increases, it can be calculated by each place value, in the same way as it has been done until now.

Want to solve

2

For the celebration of the 100th anniversary of the opening of the school, commemorative items will be distributed to 436 children. Each item has a cost of 315 yen. How much is the total cost?

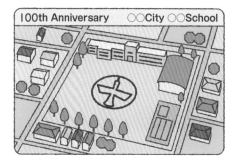

① Let's write a math expression.

② Hiroto thought the following. Let's explain Hiroto's idea.

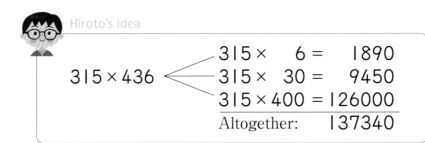

Hiroto's idea

$$315 \times 436 \begin{cases} 315 \times 6 = 1890 \\ 315 \times 30 = 9450 \\ 315 \times 400 = 126000 \end{cases}$$

Altogether: 137340

Way to see and think

Even if the number of digits increases, it can be calculated by the same way in vertical form?

③ Based on Hiroto's idea, let's think about how to calculate in vertical form.

2 I want to buy as many cans of juice as possible with **5000** yen. I went to the store and each can is sold for **68** yen. How many cans of juice can I buy?

① Let's write a math expression.

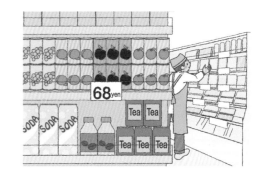

② Let's think about how to calculate in vertical form.

First, try to think approximately how many cans you can buy.

Daiki

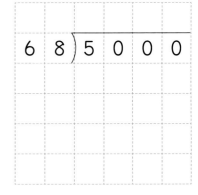

$$6\,8 \overline{\smash{)}\,5\;0\;0\;0}$$

3 Let's solve the following calculations.

① 3064 + 1987   ② 6102 − 2938

③ 738 × 952   ④ 6432 ÷ 67

What you can do now

☐ **Understanding the calculations of numbers with many digits.**

1 Let's solve the following calculations.

① $68942 + 10940$ ② $1549 + 79328$

③ $45030 + 9020$ ④ $24358 + 46852$

⑤ $45625 - 3088$ ⑥ $39510 - 10620$

⑦ $76001 - 3965$ ⑧ $47521 - 36201$

⑨ 362×204 ⑩ 429×675

⑪ 107×369 ⑫ 578×225

⑬ $9792 \div 34$ ⑭ $8624 \div 28$

⑮ $3551 \div 53$ ⑯ $5695 \div 67$

☐ **Can make a math expression and find the answer.**

2 The number of attendance in the 2017 soccer J. League was explored. The number of attendance in J1 was 5778178 people and in J2 was 3219936 people. Let's answer the following questions.

① How many people is the total number of attendance in J1 and J2 altogether?

② How many people is the difference between J1 and J2?

3 Each of the 107 students in 4th grade will make a clay ornament. 485 g of clay will be distributed to each student. How many grams of clay will be needed in total?

4 There is a 35 m ribbon that will be cut into 38 cm pieces. When this is done, how many 38 cm ribbon pieces can be made? How many centimeters will remain?

Which one is larger?

That land is for sale.

Land Ⓐ
18 million yen
14m
6m

Land Ⓑ
20 million yen
9m
10m

The price of land seems to be determined by size of area.

I think that the land with longer surrounding length is larger.

14m
6m
Land Ⓐ

9m
10m
Land Ⓑ

So, why is the price of Land Ⓐ cheaper, even when the surrounding length is longer?

Problem What kind of representation methods are there for the size of area?

13 Area
Let's explore how to represent and how to find the size of area.

1 Area

Want to solve How to represent the size of area

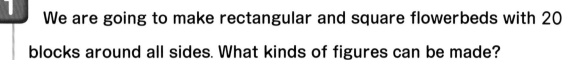

1 We are going to make rectangular and square flowerbeds with 20 blocks around all sides. What kinds of figures can be made?

① With 20 blocks, the following rectangles and square were made. Which figure is the largest?

Way to see and think

Until now, when comparing length and weight, we have used the following methods:
· comparing directly
· comparing with the trace of another thing
· comparing by number.

Ⓐ

Ⓑ

Let's try to make using the blocks in page 161.

Ⓒ 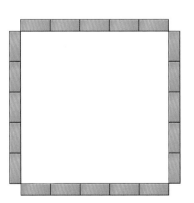   Ⓓ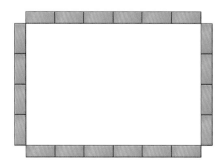

All the surrounding sides have 20 blocks, but the size of area is...

Daiki

Areas can't be compared by the surrounding length.

Yui

Purpose Can the size of area be represented by a number?

② The sizes of area Ⓒ and Ⓓ were compared as follows. Let's explain the ideas of the following children.

Nanami's idea

Place Ⓒ on top of Ⓓ, then compare the remaining parts by placing one on top of the other.

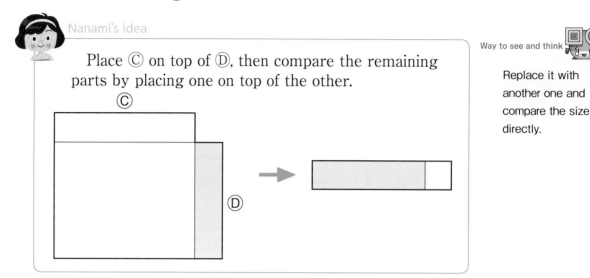

Way to see and think

Replace it with another one and compare the size directly.

Yui's idea

Draw squares whose length of the side is the same as the length of one edge of the block.

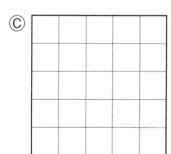 

Way to see and think

Decide one unit and compare.

Summary

The size of area can be represented by the number of unit squares.

The size of area is the amount of space surrounded by lines. The size represented by a number is called **area**.

2 There are two sheets of colored paper, Ⓐ and Ⓑ. Which one is larger and by how much?

Ⓐ

Ⓑ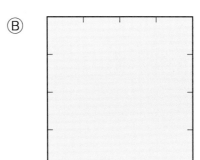

Way to see and think

What should you consider as one unit?

The area which is the same as the area of a square with a side of 1 cm is called **one square centimeter** and is written as 1 cm². The unit cm² is a unit of area.

 Let's measure the area of various things by tiling 1-cm² squares on page 161.

2 How many cm² are the following areas?

①

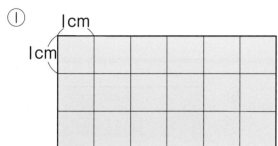

②

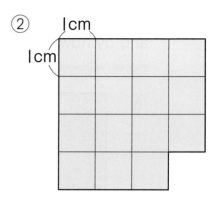

3 How many cm² is the area of the colored figures?

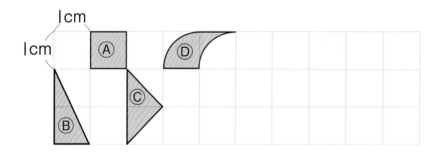

Way to see and think

If you think separately, you can understand that they all have the same area.

Let's try to draw other figures with the same area.

4 How many cm² is the area of the colored figures?

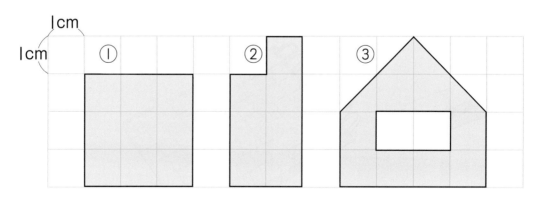

3 Let's use the grid paper to draw various figures with an area of 12 cm².

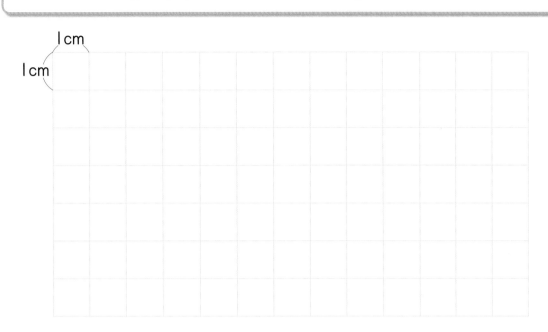

 Let's explore the following rectangles.

① As for the surrounding length, which one is longer?

② As for the area, which one is larger?

③ From ① and ②, what kind of things can you understand? Let's try to discuss.

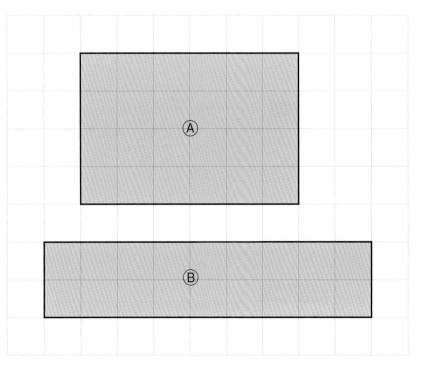

Want to think

1 Let's think about how to find the area of a rectangle with a length of 4 cm and a width of 5 cm.

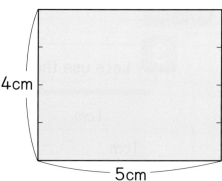

① How many 1-cm² squares can be lined up along the length?

② How many 1-cm² squares can be lined up along the width?

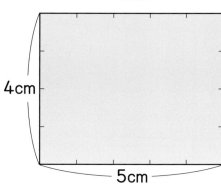

③ How many 1-cm² squares can be tiled inside the rectangle? Also, how many cm² is the area of the rectangle?

④ Let's find the area of the rectangle by using multiplication.

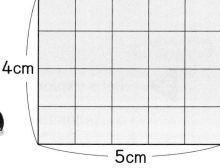

Nanami

Just count the number of 1-cm² squares.

I just need to know the number for the length and width.

Hiroto

Purpose Can the area of a rectangle be found with multiplication?

Want to explain

Number of
1-cm² squares ······ 4 × 5 = ☐

⑤ Let's explain how the math sentence shown on the right was considered.

Number of squares for length	Number of squares for width	Total number of squares
4	× 5 =	☐
Length (cm)	Width (cm)	Area (cm²)

Summary

The area of a rectangle can be found by using the length and width as shown below.

Area of a rectangle = length × width

The area of any rectangle can be found with the math sentence "area of rectangle = length × width." This math sentence is called a **formula**.

The area of a rectangle can also be found by using "width × length."

Want to try

1 How many cm² is the area of a square with a side of 3 cm? Let's think about this in the same way as with a rectangle.

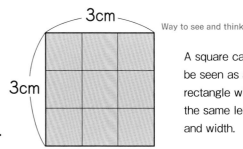

Way to see and think

A square can be seen as a rectangle with the same length and width.

The area of a square is found by using the following formula.

$$\boxed{\text{Area of a square} = \text{side} \times \text{side}}$$

Want to explore

2 Let's measure the sides of the following squares and rectangles, and find their areas.

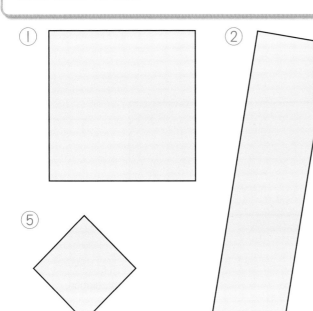

① ② ③ ④ ⑤

2 I want to make a rectangle with an area of 40 cm² and a width of 8 cm.

How many cm should the length be?

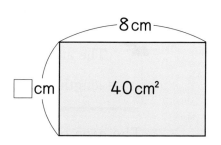

① As shown on the right, it was considered by using the formula to find the area of a rectangle. Let's explain about this idea.

$$\square \times 8 = 40$$
$$\square = 40 \div 8$$

② Let's find the length.

3 I want to make a rectangle with an area of 50 cm² and a width of 10 cm. How many cm should the length be?

Area of a combined figure

Activity

3 How many cm² is the area of the following figure?

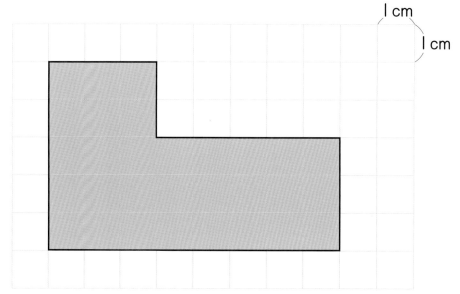

1 cm

1 cm

 We can't use the formulas because it's not a rectangle nor square.

Daiki

Let's improve the calculation, can I use the formulas?

Yui

Purpose Can we find the area of a figure that is not a rectangle nor square?

① Let's explain the ideas of the following children.

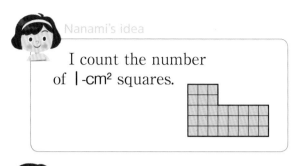

Nanami's idea

I count the number of 1-cm² squares.

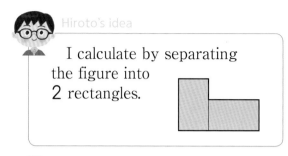

Hiroto's idea

I calculate by separating the figure into 2 rectangles.

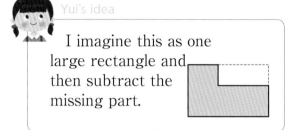

Yui's idea

I imagine this as one large rectangle and then subtract the missing part.

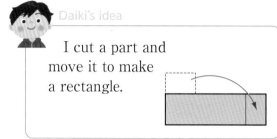

Daiki's idea

I cut a part and move it to make a rectangle.

② From the ideas in ① , which one can you use anytime?

Summary

We can find the area of figures that are not rectangles nor squares by separating them into rectangles or squares or by subtracting the missing part.

4 How many cm² is the area of the following figure?

Let's think about how to find the area.

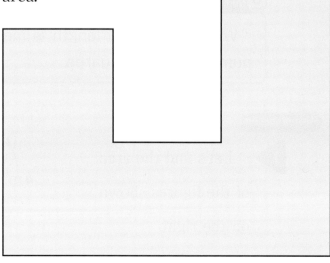

Want to think

1 The flowerbed on the right is a rectangle with a length of 3m and a width of 6m. Let's find the area of the flowerbed.

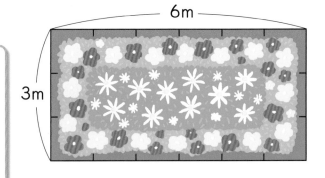

6m

3m

Nanami: If you calculate by replacing meters with cm, the number will be large.

If there are other units to represent...

Hiroto

Purpose What is the representation method of area when the unit of length is meters?

① In the flowerbed, how many squares with a side of 1m can be placed in total?

The area of a square with a side of 1m is **one square meter** and is written as 1m². The unit m² is also a unit of area just like cm².

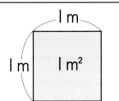

1m

1m   1m²

② How many m² is the area of the flowerbed?

Summary

When the unit of length is meters, the area can be represented by the number of 1-m² squares.

Want to confirm

1 Let's find the area of the figures shown on the right.

① 5m

4m

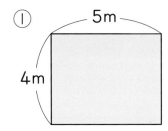

② 6m

6m

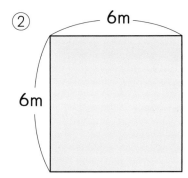

2 Let's explore how many cm² there are in 1m².

① How many 1-cm² squares can be lined up along the length? Also, how many along the width?

② How many cm² are there in 1m²?

Since 1 m = 100cm,

100 × 100 = ☐

| 1m² = 10000cm² |

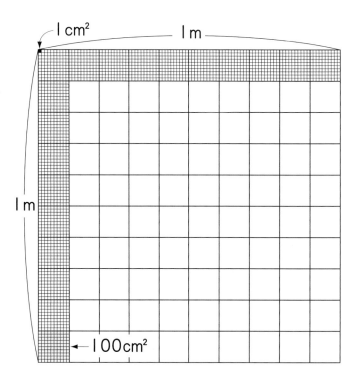

1 cm²   — 1 m —

1 m

←100cm²

 3 A newspaper poster will be made with a length of 80 cm and a width of 2 m. How many cm² is the area of the poster?

Way to see and think

To find the area, you need to align the units.

 4 Let's make a 1-m² square and explore how many people can stand on top of it.

Let's predict before trying to place people on the square.

2 Let's explore a rectangular field that has a length of 30 m and a width of 40 m.

10m

10m

30m

40m

① How many m² is the area of this field?

② How many squares with a side of 10 m can be placed in this field?

The area of a square with a side of 10 m is called **one are**, and is written as 1a. The unit a is used to represent the area of lands such as rice fields and cornfields.

③ How many a is the area of the field?

| Length of 1 side | 1 m | 10 times → | 10 m |

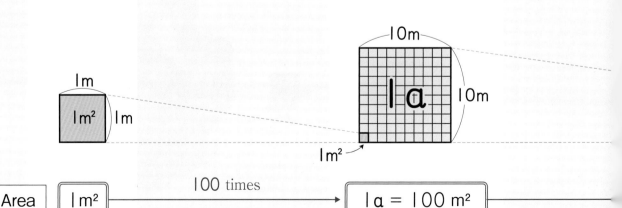

| Area | 1 m² | 100 times → | 1a = 100 m² |

5 Let's explore a square farm with a side of 600 m.

① How many m² is the area of this farm?

② How many squares with a side of 100 m can be placed in this farm?

The area of a square with a side of 100 m is called **one hectare** and is written as 1 ha. The unit ha is used to represent the area of a farm or forest.

③ How many ha is the area of the farm?

10 times ——————————→ 100 m ——————— 10 times

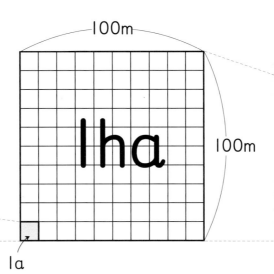

100m

1 ha

100m

1 a

100 times ——————————→ 1 ha = 10000 m² ——————— 100 times

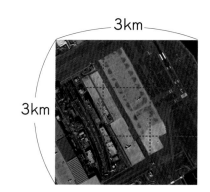

3km

3km

The photograph on the right shows an airport. This is a square with a side of **3 km.**

① How many squares with a side of l km can be placed inside this photograph?

The area of a square with a side of l km is **one square kilometer** and is written as l km². The unit km² is used to represent large areas such as islands, prefectures, and countries.

② How many km² is the area of the photograph?

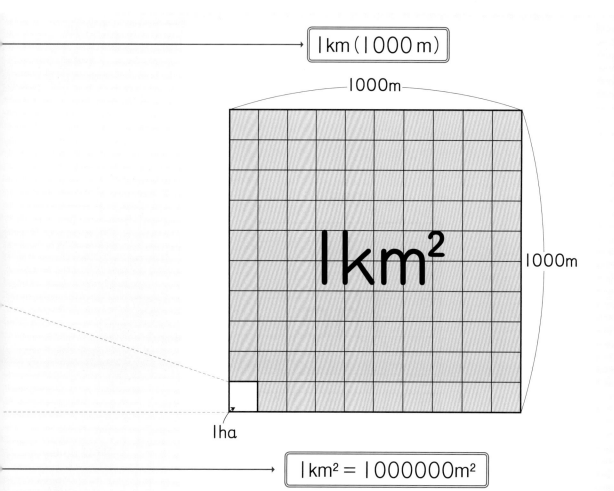

l km (l000 m)

1000m

1km²

1000m

lha

l km² = l000000 m²

1

Let's explore the relationship between the length of the side of a rectangle or square and the unit of area.

① The relationship between the length of one side and the area of a square was explored. Let's discuss what kind of relationships there are.

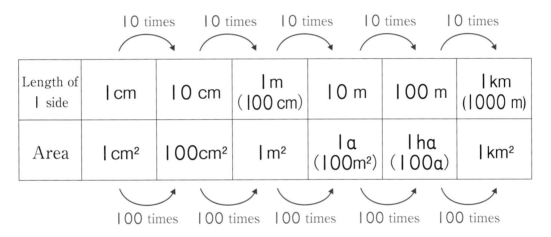

Length of 1 side	1 cm	10 cm	1 m (100 cm)	10 m	100 m	1 km (1000 m)
Area	1 cm²	100cm²	1 m²	1 a (100m²)	1 ha (100a)	1 km²

② Let's compare the area of a rectangle with a length of **2 cm** and a width of **3 cm** with the area of a rectangle with a length of **20 cm** and a width of **30 cm**.

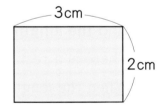

In a square or a rectangle, if each of the sides becomes 10 times longer then the area becomes 100 times larger.

③ As for length units, 1 km is 1000 times of 1 m. As for area, how many times of 1 m² is 1 km²?

What you can do now

☐ **Can find the area.**

1 Let's find the area of the following figures.

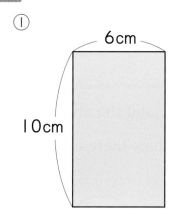

① 6cm 10cm

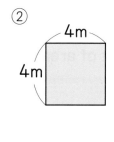

② 4m 4m

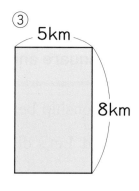

③ 5km 8km

☐ **Can improve the calculation and find the area.**

2 Let's find the area of the following figure.

5cm

8cm

3cm

12cm

☐ **Can find the length of the side by using the formulas for area.**

3 There is a rectangle with an area of 518 cm².

When the length is 14 cm, how many cm is the width?

☐ **Understanding large areas and area units.**

4 The relationship between the length of 1 side and the area of a square was explored. Let's write the number or unit that applies in the following ☐.

Length of 1 side	1 cm	10 cm	☐ m	10 m	100 m	1 km
Area	1 ☐	☐ cm²	1 m²	1 ☐ (☐ m²)	1 ha (☐ a)	1 ☐

Supplementary Problems p.150

Usefulness and efficiency of learning

1 Let's find the area of the following figures by using the unit inside the [ ].

①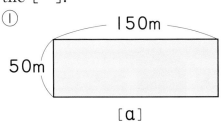

[a]

②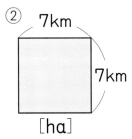

[ha]

2 Let's find the area of the colored part in the following figure.

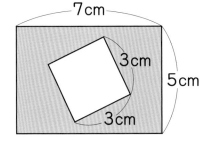

3 Let's fill in each ☐ with a number.

①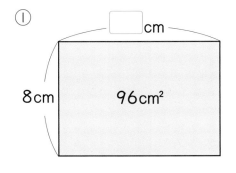

②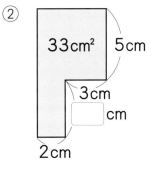

4 In the following rectangular field, the width of the path is 10 m. How many m² is the area of the field? Also, how many a is it?

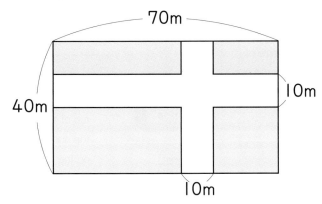

Reflect Connect

Problem

Figures A, B, and C were made with a string that is 24 cm long. Let's compare the size of A, B, and C.

A
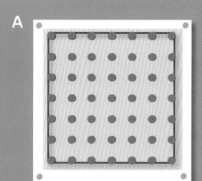

B
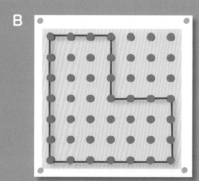

C
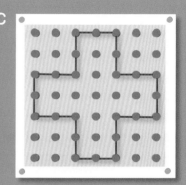

◎Comparing the size of B and C with the size of A…

> It will be clear if you place B and C on top of A.

· Just watching A, the size is larger than B and C.
· Without placing one on top of the other, it can be compared as follows.

A
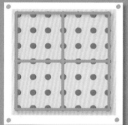

There are ☐ squares.

B
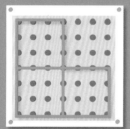

There are ☐ squares.

A
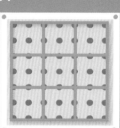

There are ☐ squares.

C
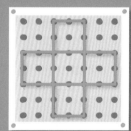

There are ☐ squares.

What kind of methods to compare size of area were there?

As for the size of area, enclose the inside with lines.

Daiki

The size of area can be compared by placing one on top of the other and counting the number of squares based on the size of one.

Nanami

◎ Comparing the size of B and C...

· Since the size of the squares in B and C are different, using cm² to represent area is easier to understand.

> The area of a square with a side that is 1 cm long.

> There are various methods.

①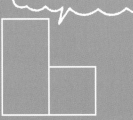

②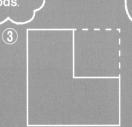

③

$3 \times 3 \times 3 =$ ☐ $6 \times 3 + 3 \times 3 =$ ☐ $6 \times 6 - 3 \times 3 =$ ☐ ☐ cm²

Let's find the area of C.
Math sentence:

> To find the area easily, which thinking method shall we use, ①, ②, or ③?

Answer: cm²

◎ A, B, and C have the same surrounding length but different sizes of area.

> How many cm² is square A?

Math sentence:

Answer: ☐ cm²

The size of area represented by a number is called area. The units of area are cm² and m². Also a, ha, and km² are used.

Yui

Also, for the area of figures like B and C, we can improve the calculation and use various methods to find the area.

Hiroto

Want to connect

Yui

There are formulas to find the area of rectangles and squares. Are there also formulas for parallelograms and other quadrilaterals?

How many liters are there in total?

Problem Can we also multiply decimal numbers in the same way as whole numbers?

14 Let's Think about How to Calculate

Let's think about judicious ways for how to calculate with decimal numbers.

1 Decimal number × whole number

Want to solve

Activity

1

There are 3 bottles that contain ☐ L of juice. How many liters are there in total?

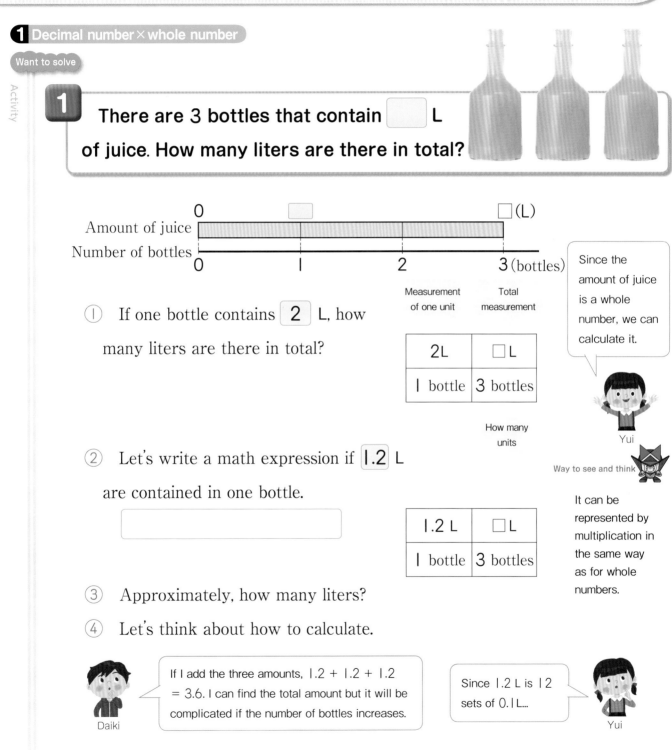

Amount of juice
Number of bottles

0   ☐   ☐(L)

0   1   2   3 (bottles)

① If one bottle contains **2** L, how many liters are there in total?

Measurement of one unit	Total measurement
2L	☐ L
1 bottle	3 bottles

How many units

Since the amount of juice is a whole number, we can calculate it.

Yui

② Let's write a math expression if **1.2** L are contained in one bottle.

1.2 L	☐ L
1 bottle	3 bottles

Way to see and think

It can be represented by multiplication in the same way as for whole numbers.

③ Approximately, how many liters?

④ Let's think about how to calculate.

Daiki: If I add the three amounts, 1.2 + 1.2 + 1.2 = 3.6. I can find the total amount but it will be complicated if the number of bottles increases.

Yui: Since 1.2 L is 12 sets of 0.1 L...

Purpose When multiplying a decimal number by a whole number, how should it be calculated?

⑤ Let's explain the ideas of the following children.

 Hiroto's idea

If I change L to dL, 1.2 L = 12 dL.

12 × 3 = 36

36 dL = ☐ L

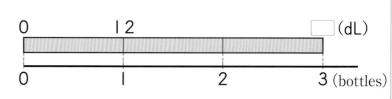

 Yui's idea

If I use 0.1 as a unit,

1.2 is 12 sets of 0.1.

12 × 3 = 36

36 sets of 0.1 is ☐.

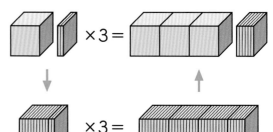

12 sets of 0.1   36 sets of 0.1

 Nanami's idea

I use the structure of decimal numbers and the rules of multiplication.

1.2 × 3 = ☐

↓ 10 times   ↑ $\frac{1}{10}$

12 × 3 = 36

In multiplication, if the multiplier or the multiplicand is multiplied by 10, the answer is also multiplied by 10.

Summary

When decimal numbers are changed into whole numbers, the multiplication of decimal numbers can be calculated in the same way as the multiplication of whole numbers.

 There are 3 bottles that contain 1.5 L of juice. How many liters are there in total?

1 If we equally divide ☐ L of juice into three bottles, how many liters will each bottle contain?

Amount of juice
Number of bottles

0 ☐ ☐ (L)

0   1   2   3 (bottles)

① If there are **6** L, how many liters of juice will there be in each bottle?

Measurement of one unit	Total measurement
☐ L	6 L
1 bottle	3 bottles

How many units

The answer can be found by "measurement of one unit = total measurement ÷ how many units."

② Let's write the math expression when there are **5.4** L of juice.

☐ L	5.4 L
1 bottle	3 bottles

Way to see and think

It can be represented by division in the same way as for whole numbers.

③ Approximately, how many liters?

④ Let's think about how to calculate.

Can I think of changing L into dL?

Yui

Can I think based on the division of whole numbers?

Hiroto

Purpose When dividing a decimal number by a whole number, how should it be calculated?

⑤ Let's explain the ideas of the following children.

Hiroto's idea

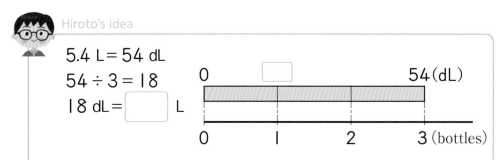

5.4 L = 54 dL
$54 \div 3 = 18$
18 dL = ☐ L

0   ☐   54 (dL)

0   1   2   3 (bottles)

Yui's idea

5.4 is 54 sets of 0.1.
$54 \div 3 = 18$
18 sets of 0.1 is ☐.

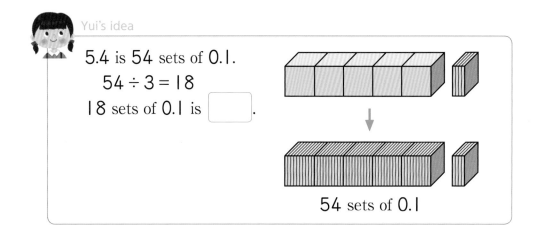

54 sets of 0.1

Nanami's idea

I used the structure of decimal numbers and the rules of division.

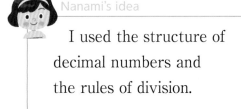

$5.4 \div 3 =$ ☐

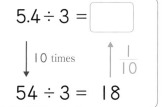

10 times $\frac{1}{10}$

$54 \div 3 = 18$

In division, if the dividend is multiplied by 10, the answer is also multiplied by 10.

Summary

    When decimal numbers are changed into whole numbers, the division of decimal numbers can be calculated in the same way as the division of whole numbers.

1   If we want to equally divide 5.1 L of juice into three bottles, how many liters will each bottle contain?

15 Multiplication and Division of Decimal Numbers
Let's think about how to multiply and divide decimal numbers in vertical form.

❶ Calculation of decimal number × whole number

Want to solve Multiplication of decimal numbers in vertical form

1

A wire is 1 m long and weighs 2.3 g.
How many grams is the weight of 4 m of this wire?

```
         0        2.3                          □(g)
Weight  [                                         ]
Length  |---------|---------|---------|---------|
         0        1         2         3         4 (m)
```

① Let's write a math expression.

② Approximately, how many grams?

Measurement of one unit	Total measurement
2.3 g	□ g
1 m	4 m

How many units

Way to see and think

You should think of the weight of 1 m as one unit.

Want to think

③ Let's think about how to calculate.

I can find it by thinking how many sets of 0.1 there are or using the rules of multiplication.
Daiki

Can I use vertical form in the same way as with whole numbers?
Yui

♈ Purpose Can we also calculate the multiplication of decimal numbers in vertical form?

④ Let's think about how to calculate in vertical form.

```
      2.3
  ×     4
```

What should I do with the decimal point?

Nanami

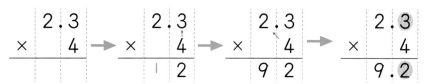

	2.3
×	4

→

	2.3 ↑
×	4
	1 2

→

	2.3
×	4
9	2

→

	2.**3**
×	4
9 .	**2**

···The number of digits after the decimal point is 1.

···The number of digits after the decimal point is 1.

Align and write from the right.

Multiply in the same way as in multiplication of whole numbers in vertical form.

The number of digits after the decimal point in the product must be the same as that in the multiplicand.

Want to think   Calculation of area

 1 How many m² is the area of a flowerbed with a length of **2.6** m and a width of **3** m?

① Let's write a math expression.

② Let's calculate in vertical form.

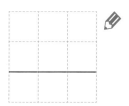

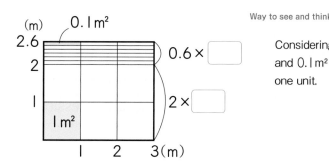

Way to see and think

Considering 1 m² and 0.1 m² as one unit.

6 sets of 1 m² are ☐ m²

18 sets of 0.1 m² are ☐ m²

Altogether: ☐ m²

Want to try

 2 Let's think about the following multiplication algorithm in vertical form.

① 0.4 × 8

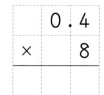

② 0.8 × 7

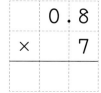

Want to confirm

 3 Let's solve the following calculations in vertical form.

① 3.2 × 3   ② 3.3 × 3   ③ 1.8 × 2

④ 2.4 × 4   ⑤ 4.3 × 6   ⑥ 0.7 × 6

2 Let's explain the following multiplication algorithm in vertical form.

In addition and subtraction, what happened with a 0 after the decimal point?

Hiroto

① 3.5 × 6　　② 0.4 × 5　　③ 2.5 × 4

```
    3 . 5          0 . 4          2 . 5
×       6      ×       5      ×       4
```

 Let's solve the following calculations in vertical form.

① 1.5 × 6　　② 4.5 × 4　　③ 2.5 × 8　　④ 0.6 × 5

Multiplications with a 2-digit multiplier

 Let's think about the following multiplication algorithm in vertical form.

① 1.6 × 14

```
    1 . 6
×   1   4
```

② 1.5 × 18

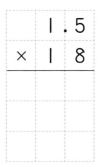
```
    1 . 5
×   1   8
```

 Let's solve the following calculations in vertical form.

① 2.2 × 12　　② 1.9 × 14　　③ 1.7 × 15　　④ 3.4 × 12

⑤ 4.8 × 21　　⑥ 0.2 × 13　　⑦ 3.5 × 18　　⑧ 2.9 × 30

 There are 13 bottles that contain 1.2 L of juice. How many liters are there in total?

3 A park has a 2.35 km path around it. If you travel 3 laps around the path by bicycle, how many kilometers would you travel in total?

① Let's write a math expression.

2.35 km	□ km
1 lap	3 laps

② Let's think about how to calculate.

By what number should 2.35 be multiplied such that it becomes a whole number?

Daiki

If I consider 0.01 as a unit, how many sets are there in 2.35?

Yui

③ Let's calculate in vertical form.

```
    2 . 3  5
×          3
```

8 Let's think about the following multiplication algorithm in vertical form.

① 0.24 × 4

② 0.04 × 5

9 A bar is 1 m long and weighs 1.25 kg. How many kilograms is the weight of 23 m of this bar?

10 Let's solve the following calculations in vertical form.

① 1.87 × 5   ② 0.63 × 5   ③ 0.23 × 4   ④ 0.12 × 7

⑤ 0.08 × 5   ⑥ 0.15 × 6   ⑦ 3.14 × 12   ⑧ 0.57 × 34

Want to solve   Division of decimal numbers in vertical form

1   If you equally divide a 5.7 m ribbon among 3 children, how many meters will each child receive?

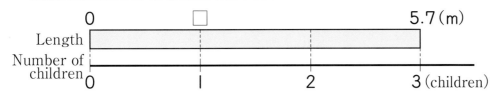

①   Let's write a math expression.

	Measurement of one unit	Total measurement
	□ m	5.7 m
	1 child	3 children

②   Approximately, how many meters?

③   Let's think about how to calculate.

How many units

I should think about how many sets of 0.1 this is.

Nanami

I can use the rules of division, but for vertical form...

Hiroto

④   Let's think about the division algorithm in vertical form.

$$3\overline{)5.7}$$

It looks like we can divide in the same way as dividing whole numbers in vertical form, but how shall we treat the decimal point in the quotient?

Yui

Division algorithm to calculate 5.7 ÷ 3 in vertical form

$$3\overline{)5.7}$$ → $$3\overline{)5.7}$$ → $$\begin{array}{r} 1.9 \\ 3\overline{)5.7} \\ 3 \\ \hline 2\ 7 \\ 2\ 7 \\ \hline 0 \end{array}$$

Align the decimal point of the quotient with the same place as that of the dividend.

When 5 is divided by 3, the quotient is written in the ones place.

As for this 27, it means 27 sets of what?

Then, calculate as if this is division of whole numbers.

Want to confirm

 1 Let's solve the following calculations in vertical form.

① 7.5 ÷ 5   ② 6.8 ÷ 2   ③ 8.4 ÷ 7

Want to think Calculation of area

 2 Let's find the length of a rectangle that has an area of 38.4 cm² and a width of 12 cm.

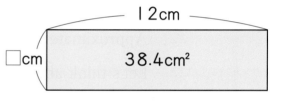
12cm

☐cm   38.4cm²

① Let's write a math expression.

Way to see and think

② Let's think about the division algorithm in vertical form.

Let's try to think how many sets of 0.1 this is.

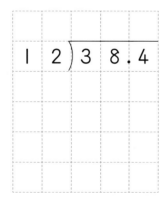

$$12\overline{)38.4}$$

Want to confirm

 3 Let's solve the following calculations in vertical form.

① 54.4 ÷ 34   ② 91.2 ÷ 24   ③ 52.9 ÷ 23

2 If you equally divide a 4.5 m tape among 9 children, how many meters will each child receive?

① Let's write a math expression. [   ]

② Hiroto thought by using vertical form as shown on the right. Let's explain Hiroto's idea.

Hiroto's idea

$$9\overline{)4.5} \rightarrow \quad (1)\quad 9\overline{)4.5}^{\;0.} \rightarrow \quad (2)\quad 9\overline{)4.5}^{\;0.5}$$
$$\begin{array}{r} 4\,5 \\ \hline 0 \end{array}$$

4 Let's solve the following calculations in vertical form.

① 3.5 ÷ 5   ② 4.8 ÷ 6   ③ 5.4 ÷ 9   ④ 0.9 ÷ 3

5 Let's explain the division algorithm for 1.61 ÷ 7 in vertical form.

$$7\overline{)1.61} \rightarrow 7\overline{)1.61}^{\;0.} \rightarrow 7\overline{)1.61}^{\;0.2} \rightarrow 7\overline{)1.61}^{\;0.23}$$

$$\begin{array}{r} 1\,4 \\ \hline 2\,1 \end{array} \qquad \begin{array}{r} 1\,4 \\ \hline 2\,1 \\ 2\,1 \\ \hline 0 \end{array}$$

6 Let's solve the following calculations in vertical form.

① 1.62 ÷ 3   ② 3.96 ÷ 4   ③ 2.25 ÷ 15   ④ 0.52 ÷ 4

Want to solve   Dividing continuously

1   If you equally divide a 7.3 m ribbon among 5 children, how many meters will each child receive?

①   Let's write a math expression.

□ m	7.3 m
1 child	5 children

②   Let's explain the division algorithm in vertical form.

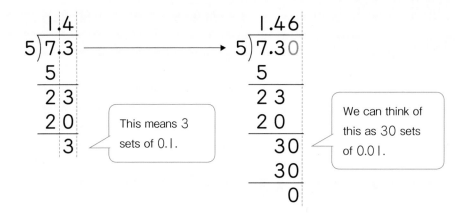

This means 3 sets of 0.1.

We can think of this as 30 sets of 0.01.

A division that is continued until the remainder is 0 is called "**dividing continuously**."

Want to try

Let's divide continuously to find the answer of $6 \div 8$.

In the case of whole number ÷ whole number, if you divide continuously, the quotient may become a decimal number.

```
    0.7
8)6.0
    5 6
      4
```

Want to confirm

Let's solve the following calculations by dividing continuously.

① $9.4 \div 4$ ② $8.6 \div 5$ ③ $7 \div 5$ ④ $5 \div 8$

2 If you equally divide 2.3 L of juice among 6 children, how many liters will each child receive?

① Let's write a math expression.

□ L	2.3 L
I child	6 children

② As shown on the right, the calculation was done by dividing continuously. How should the answer be given?

Even if I divide continuously, the remainder is not divisible by the divisor.

Daiki

0.3833..., it will continue on and on.

Yui

```
     0.3833
6)2.3
    18
    ‾‾‾
     50
     48
     ‾‾
      20
      18
      ‾‾
       20
       18
       ‾‾
        2
```

③ As for the quotient, let's find the nearest tenths place round number by rounding off the hundredths place.

When the remainder is not divisible by the divisor and the number of digits becomes too long, the quotient can be rounded.

3 If you divide a 16.3 m string into 3 equal sections, about how many meters will each be? As for the quotient, let's find the nearest tenths place round number by rounding off the hundredths place.

4 Let's solve the following calculations. As for the quotient, let's find the nearest tenths place round number by rounding off the hundredths place.

① 5.5 ÷ 8   ② 9.9 ÷ 7   ③ 67.8 ÷ 79

3 There is a 13.5 m tape. One floral decoration is made with 2 m of tape. How many floral decorations can be made and how many meters will remain?

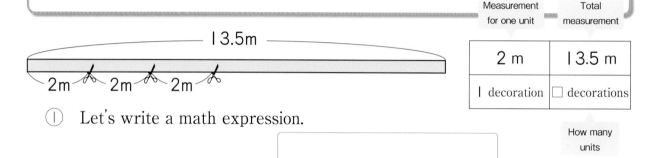

Measurement for one unit	Total measurement
2 m	13.5 m
1 decoration	☐ decorations

① Let's write a math expression.

How many units

② The calculation is shown on the right. How many meters is the remainder?

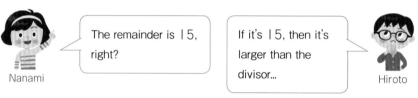

Nanami: The remainder is 15, right?

Hiroto: If it's 15, then it's larger than the divisor...

```
      6.
   ────────
 2)1 3.5
   1 2
   ────────
     1 5
```

🌱Purpose In division of decimal numbers, how should the remainder be represented?

③ In the calculation shown in ②, 15 represents 15 sets of what?

④ Let's find the remainder and confirm the answer.

Dividend = Divisor × Quotient + Remainder

13.5 =   2   ×   6   + ☐

💡 Summary

In division of decimal numbers, the decimal point of the remainder is aligned with the same place as the original point of the dividend.

```
      6.
   ────────
 2)1 3.5
   1 2
   ────────
     1˙5
```

5 ▶ When 47.6 dL of water are divided into 3 dL cups, how many cups will be filled? How many dL will remain?

Until which place should we divide continuously?

1 There are 3 bottles and each contains 1.5 L of juice. How many liters of juice are there in total?

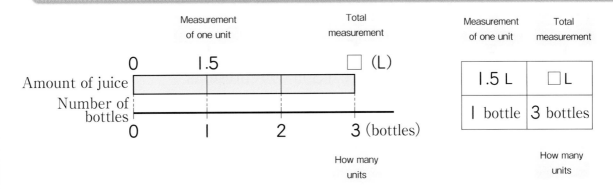

1 There are 6 plates with the same weight. The total weight is 5.1 kg. How many kilograms does each plate weigh?

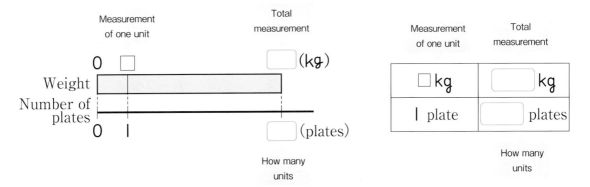

2 A wire is 1 m long and weighs 3 g. How many meters is the length of this wire when it weighs 28.5 g?

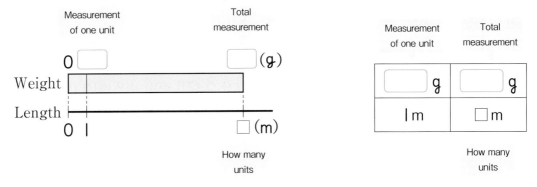

What you can do now

☐ **Understanding the structure of decimal number × whole number and decimal number ÷ whole number.**

1 Let's summarize how to multiply and divide decimal numbers.

① As for 2.7×5, if you consider how many sets of ☐ it has, then it becomes $27 \times 5 = 135$. So, the answer to 2.7×5 is ☐.

② As for $6.48 \div 9$, if you consider how many sets of ☐ it has, then it becomes $648 \div 9 = 72$. So, the answer to $6.48 \div 9$ is ☐.

③ As shown on the right, since 13 in ⓐ represents 13 sets of ☐, then $9.3 \div 4 = 2$ remainder ☐.

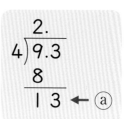

☐ **Can calculate decimal number × whole number and decimal number ÷ whole number.**

2 Let's solve the following calculations in vertical form.

① 2.4×3 ② 6.8×4 ③ 5.3×7

④ 0.6×7 ⑤ 2.8×12 ⑥ 3.7×15

⑦ 9.2×49 ⑧ 70.5×73 ⑨ 6.52×4

⑩ 0.26×8 ⑪ 0.46×5 ⑫ 0.38×18

⑬ $7.2 \div 4$ ⑭ $9.6 \div 6$ ⑮ $6.5 \div 5$

⑯ $12.6 \div 7$ ⑰ $41.6 \div 26$ ⑱ $20.8 \div 13$

⑲ $49.4 \div 19$ ⑳ $65.61 \div 27$ ㉑ $8.1 \div 9$

㉒ $0.8 \div 2$ ㉓ $15.36 \div 32$ ㉔ $0.81 \div 3$

☐ **Can make a math expression and find the answer.**

3 Let's answer the following problems.

① There is a book with a length of 14.8 cm and a width of 21 cm. How many cm² is the area of the cover of this book?

② There is a rectangular flowerbed with an area of 17.1 m². The length of this flowerbed is 3 m. Let's find the width of the flowerbed.

Supplementary Problems
p.152

Usefulness and efficiency of learning

1   Let's solve the following calculations. As for the quotient, let's find the nearest tenths place round number by rounding off the hundredths place.

①   2.63 ÷ 3   ②   40.4 ÷ 6

③   30.42 ÷ 14   ④   5.6 ÷ 39

☐ Understanding the structure of decimal number ÷whole number.

☐ Can calculate decimal number ÷whole number.

2   There are 9 L of rice that weigh 8 kg. How many kilograms does 1 L of rice weigh? As for the answer, let's find the nearest tenths place round number by rounding off the hundredths place.

☐ Can make a math expression and find the anwer.

3   There are 25 books and each weighs 0.14 kg. How many kilograms is the total weight?

☐ Can make a math expression and find the anwer.

4   If a 36.5 m rope is equally divided into 5 pieces, how many meters is each piece? Also, if the rope is cut into pieces that are 5 m long, how many pieces can be cut? How many meters will remain?

☐ Can make a math expression and find the anwer.

5   A wire is 1 m long and weighs 8 g. How many meters is the length of this wire when it weighs 36.8 g?

☐ Can make a math expression and find the anwer.

6   A land with a length of 9.5 m and a width of 12 m was divided into Ⓐ and Ⓑ as shown on the right.

   Let's find the number that applies in ☐ when the area of Ⓐ and Ⓑ is the same.

☐ Can make a math expression and find the anwer.

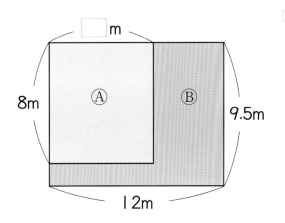

How many times:

Try Boccia.

Boccia is a sport in which a ball is thrown or rolled to approach its target. Himari and her friends had the personal experience to play Boccia.

Want to explore

1 The table on the right summarizes balls' distance leading up to the target. Let's explore the relationship between each record and Himari's record.

Distance to target

	Distance (cm)
Himari	8
Arata	16
Riko	24
Shoma	20
Yuzuki	12

① How many times of Himari's record is Arata's record? Also, how many times of Himari's record is Riko's record?

Riko [_____] 24cm   24 ÷ 8 = [ ] (times)

Arata [_____] 16cm   16 ÷ 8 = [ ] (times)

Himari [_____] 8cm

0   1   [ ]   [ ] (times)

Way to see and think

It was total amount ÷ amount of one unit = how many units.

② As for Shoma's record, can you answer how many times of Himari's record it is?

16 20cm

Shoma [_____]

Himari [_____] 8cm   20 ÷ 8 = [ ] (times)

0   1   2   [ ]   3 (times)

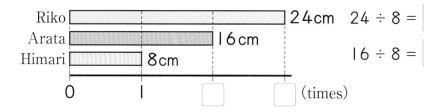

Amount of one unit   Total amount

8 cm	20 cm
1 time	☐ times

Times

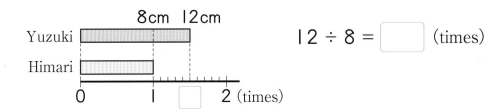

When you consider
Himari's record as
one unit, then you
think about how
many units
Shoma's record is.

When you consider Himari's record as 1, then Shoma's record is 2.5 times.

Such as "2.5 times," decimal numbers are also used to represent "how many times of." 2.5 times means that when 8 cm is considered as 1, then 20 cm represents 2.5.

Want to confirm

③ How many times of Himari's record is Yuzuki's record?

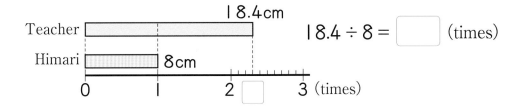

$12 \div 8 = \boxed{}$ (times)

Want to try

④ The teacher's record was 18.4 cm.

How many times of Himari's record is the teacher's record?

Teacher

Himari 8cm

18.4cm

$18.4 \div 8 = \boxed{}$ (times)

0   1   2 ☐ 3 (times)

Utilizing rule of three on a 4-cell table

1　A wire is | m long and weighs 5.8 g. How many grams is the weight of 5 m of this wire?

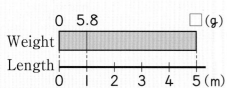

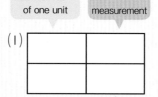

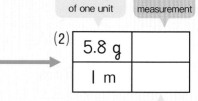

> **How to create the three on a 4-cell table.**
>
> (|) Write a table with 4 entries.
>
> (2) Since | m of wire weighs 5.8 g, write " | m" and "5.8 g" in the left column.
>
> (3) Since you do not know the weight for 5 m, write "5 m" and "□ g" in the right column.

Even if you write it in line with the first figure, you can also write as shown on the right.

	m	5 m
5.8 g	□ g

2　A wire is 6 m long and weighs 25.2 g. How many grams is the weight of | m of this wire?

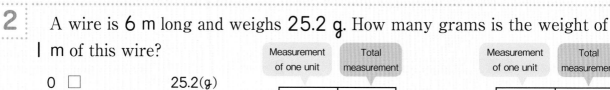

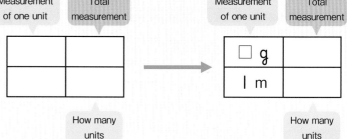

3　A wire is | m long and weighs 3 g. How many meters is the length of this wire when it weighs 46.5 g?

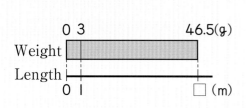

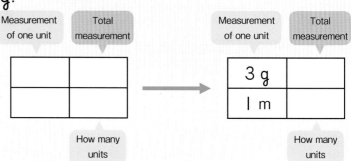

1

(3)

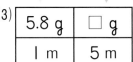

Measurement of one unit	Total measurement
5.8 g	□ g
1 m	5 m

How many units

The same unit is placed in the horizontal row.

| Measurement of one unit | × | How many units | = | Total measurement |

So,

$$5.8 \times 5 = 29$$

Answer: 29 g

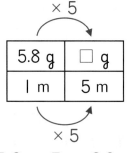

× 5

5.8 g	□ g
1 m	5 m

× 5

$$5.8 \times 5 = 29$$

Answer: 29 g

2

Measurement of one unit	Total measurement
□ g	25.2 g
1 m	6 m

How many units

| Total measurement | ÷ | How many units | = | Measurement of one unit |

So,

$$25.2 \div 6 = 4.2$$

Answer: 4.2 g

× 6

□ g	25.2 g
1 m	6 m

× 6

$$\square \times 6 = 25.2$$
$$25.2 \div 6 = 4.2$$

Answer: 4.2 g

3

Measurement of one unit	Total measurement
3 g	46.5 g
1 m	□ m

How many units

| Total measurement | ÷ | Measurement of one unit | = | How many units |

So,

$$46.5 \div 3 = 15.5$$

Answer: 15.5 m

×□

3 g	46.5 g
1 m	□ m

×□

$$3 \times \square = 46.5$$
$$46.5 \div 3 = 15.5$$

Answer: 15.5 m

Active Learning!!

Do you know how a math sentence is made?

Want to think   Materials for a medal

At Yui's school, children who have not yet entered Elementary School were invited to a study meeting.

It is scheduled that **24** children will participate, so a medal will be made for each of them.

The materials to create a medal for one child is shown on the right.

> **Materials for a medal**
> - 80cm ribbon
> - Cardboard cut into a circle

The teacher has provided a **2000** cm ribbon and a rectangular cardboard that has a length of **39** cm and a width of **54** cm. Nanami, Daiki, and Hiroto made the following math sentences to consider whether or not the ribbon provided by the teacher is enough.

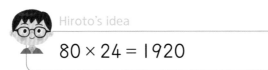

Hiroto's idea
$$80 \times 24 = 1920$$

Daiki's idea
$$2000 \div 80 = 25$$

Nanami's idea
$$2000 \div 24 = 83.3\ldots$$

1 In the math sentence made by Hiroto, what do the multiplicand and multiplier represent? Also, let's explain how he made the sentence.

2 Let's discuss the similarities and differences between the math sentences made by Daiki and Nanami.

3 Based on the results from the math sentences made by the **3** children, let's explain whether or not the ribbon is enough.

4 Daiki explained, as shown below, that he could cut **24** squares with a side of **9** cm from the rectangular cardboard that has a length of **39** cm and a width of **54** cm.

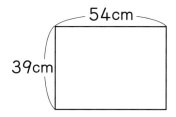

Let's write the continuation of Daiki's explanation by using words and math expressions.

Daiki's explanation

The cardboard has a width of **54** cm.
Since one side of the square is **9** cm, $54 \div 9 = 6$,
I can write **6** squares in the width.

54cm

01405

Let's think about how to represent and calculate numbers.

In Japanese abacus, one unit point is determined as the ones place. Then from the ones place to the left, the tens place, hundreds place, etc. are defined. To the right, the tenths place, hundredths place, etc. are defined. Also, the thousands place, millions place, etc. become unit points to the left of the ones place.

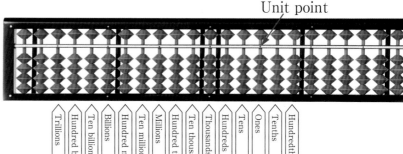

Unit point

Trillions | Hundred billions | Ten billions | Billions | Hundred millions | Ten millions | Millions | Hundred thousands | Ten thousands | Thousands | Hundreds | Tens | Ones | Tenths | Hundredths

1 How to represent numbers

Want to think

1 Let's read the following numbers.

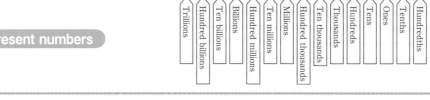

Ones place   Ones place

①   ②

Ones place   Ones place

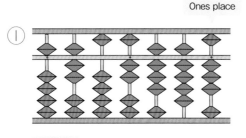
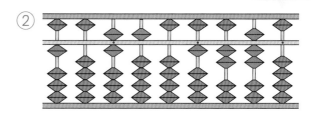

③   ④

Way to see and think

Want to confirm

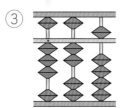

1 Let's represent the following numbers on the abacus after setting the ones place.

① 8126   ② 1375604   ③ 1200000000

④ 12.9   ⑤ 0.8   ⑥ 0.16

Large numbers such as 万 (ten thousand), 億 (one hundred million) or 兆 (trillion) are used to read every 4 digits.

Want to know   Addition using an abacus

Let's solve the following calculations.

Ones place

① 58 + 54

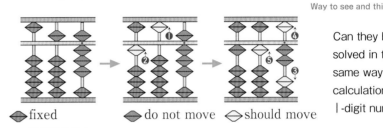

◆fixed   ◆do not move   ◇should move

Way to see and think

Can they be solved in the same way as calculations of 1-digit numbers?

② 4.8 + 2.3   Ones place

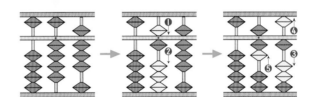

③ 0.36 + 0.47   Ones place

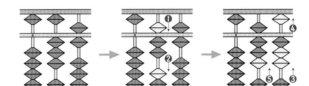

④ 5 billion
　+ 1 billion

Billions place

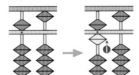

Want to confirm

Let's solve the following calculations.

① 67 + 54　　　83 + 39　　　94 + 28　　　36 + 89

　83 + 691　　78 + 72　　　58 + 93　　　76 + 79

② 0.3 + 7.5　　2.8 + 1.4　　0.1 + 0.9　　1.4 + 3.7

　0.48 + 0.25　4.36 + 2.89　5.72 + 3.45

③ 4 billion + 7 billion　60 billion + 90 billion

　40 trillion + 50 trillion

2   Let's solve the following calculations.

①   112 − 54

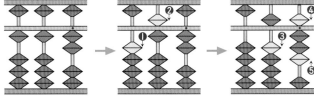

②   144 − 76

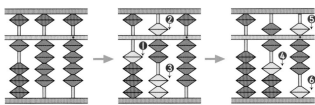

③   3.3 − 1.5

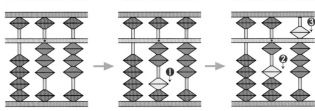

④   0.58 − 0.24

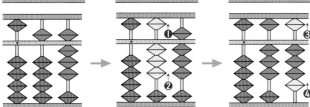

⑤   7 billion − 4 billion

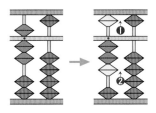

3   Let's solve the following calculations.

①   165 − 88   102 − 29   156 − 89   104 − 25
     123 − 67   143 − 66   134 − 78   121 − 76

②   2.9 − 0.4   8.3 − 0.5   3.7 − 1.7   12.6 − 3.9
     0.72 − 0.45   3.41 − 1.63   6.38 − 2.29

③   8 billion − 2 billion   700 trillion − 600 trillion   10 billion − 3 billion

How long is the surrounding length of a tree?

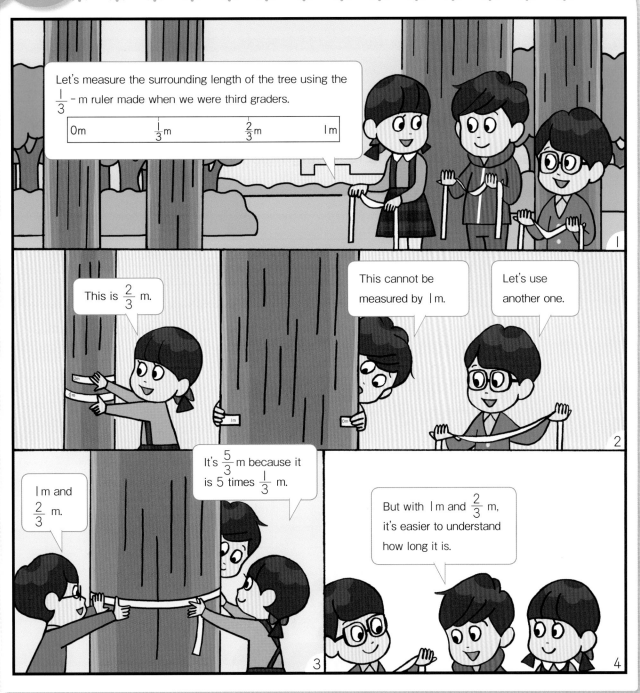

Problem How can we represent fractions larger than 1?

17 Fractions

Let's think about the size of fractions and how to calculate.

1 Fractions larger than |

Want to represent

1 When measuring the surrounding length of a tree with a $\frac{1}{3}$-m ruler, the trees Ⓐ and Ⓑ had the following tape lengths. How many meters does each tape represent?

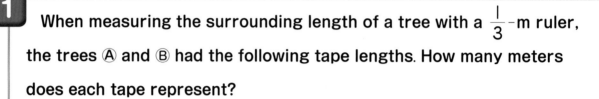

① How many meters is the length of Ⓐ?

② How many meters is the length of Ⓑ?

Way to see and think

What should we consider as one unit?

Let's represent these in the following two ways.

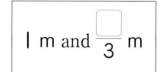
□ sets of $\frac{1}{3}$ m are $\frac{□}{3}$ m

| m and $\frac{□}{3}$ m

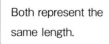
Both represent the same length.

The representation methods are different.

⚇ **Purpose** How can we represent fractions larger than |?

The sum of 1 m and $\dfrac{2}{3}$ m is written as $1\dfrac{2}{3}$ m and is read as "**one and two thirds meters.**"

$1\dfrac{2}{3}$ m indicates the same length as $\dfrac{5}{3}$ m.

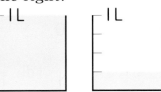

$$1\dfrac{2}{3} = \dfrac{5}{3}$$

Want to try

1 How many liters is the amount of water shown on the right?

① 1 L and how many liters more?

1 L and $\dfrac{\square}{\square}$ L ➡ $1\dfrac{\square}{\square}$ L

② If you think as shown on the right, how many sets of $\dfrac{1}{4}$ L are there?

$\dfrac{\square}{4}$ L

Fractions in which the numerator is smaller than the denominator like $\dfrac{2}{3}$ and $\dfrac{1}{4}$ are called **proper fractions.** Fractions in which the numerator is equal to or larger than the denominator like $\dfrac{4}{4}$ and $\dfrac{7}{4}$ are called **improper fractions.**

Fractions which are the sum of a whole number and a proper fraction like $1\dfrac{2}{3}$ and $1\dfrac{1}{4}$ are called **mixed fractions.**

🔎 Summary

Fractions larger than 1 can be represented as mixed fractions or improper fractions.

2 Let's represent the following lengths and amounts of water as mixed fractions and improper fractions.

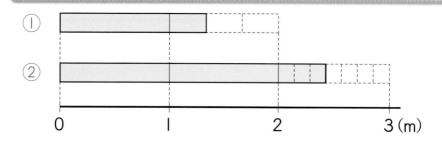

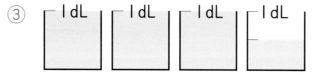
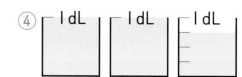

2 Let's represent 5 sets, 6 sets, 7 sets, and 8 sets of $\frac{1}{5}$ m as mixed fractions and improper fractions with the tape diagram and number line.

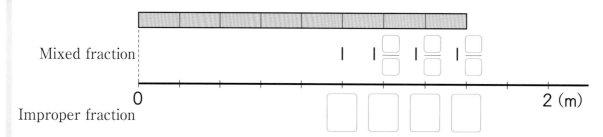

Mixed fraction

Improper fraction

Proper fractions are smaller than 1, mixed fractions are larger than 1, and improper fractions are equal to 1 or larger than 1.

3 What numbers are represented by arrows Ⓐ~Ⓓ on the following number line? Let's write the fractions larger than 1 as mixed fractions and improper fractions.

Activity

3 Let's change $2\frac{4}{5}$ to an improper fraction.

① Let's think by using the diagram on the right.

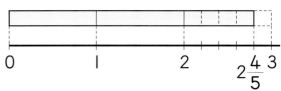

② Let's explain the ideas of the two children.

 Daiki's idea

Thinking about fractions with 5 in the denominator,

$2\frac{4}{5}$ is $\frac{5}{5}$, $\frac{5}{5}$, and $\frac{4}{5}$.

$2\frac{4}{5} = \frac{\square}{5}$

Yui's idea

Considering $\frac{1}{5}$ as a unit, we have $\square$ sets of $\frac{1}{5}$, calculating $5 \times 2 + 4$.

$2\frac{4}{5} = \frac{\square}{5}$

Want to try

 4 Let's change $\frac{7}{4}$ to a mixed fraction.

$\frac{7}{4}$ is the sum of $\frac{4}{4}$ and $\frac{3}{4}$.

Since $\frac{4}{4}$ is 1, then we have $\frac{7}{4} = \square\frac{\square}{4}$

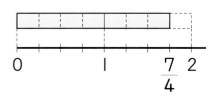

5 Let's change $\frac{15}{5}$ to a whole number.

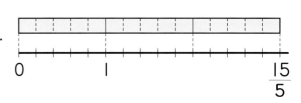

Want to confirm

 6 Let's change mixed fractions to improper fractions, and improper fractions to mixed fractions or whole numbers.

① $4\frac{2}{3}$ ② $2\frac{1}{6}$ ③ $\frac{13}{4}$ ④ $\frac{9}{5}$ ⑤ $\frac{8}{2}$

Want to explore

1

Let's investigate the size of fractions by using the following number lines.

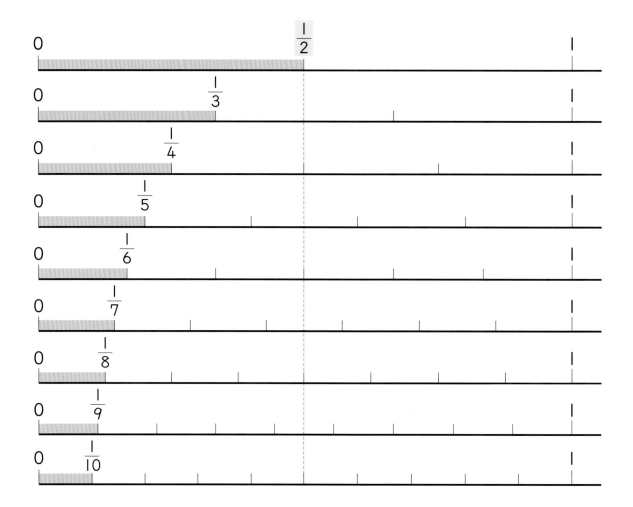

Ⓨ Purpose   Does the size of a fraction have any relationship to the size of the denominator and the numerator?

① Let's read $\frac{1}{2}$, $\frac{1}{3}$, $\frac{1}{4}$, $\frac{1}{5}$, $\frac{1}{6}$, $\frac{1}{7}$, $\frac{1}{8}$, $\frac{1}{9}$, and $\frac{1}{10}$ from smallest to largest.

② Let's replace the numerators in ① with a **2**, write the fractions in the above number line, and read them from smallest to largest.

③ Let's refer to the number line on the previous page and write the fractions that are equivalent to the following fractions.

Way to see and think

Looking at the number lines together, there are fractions that line up vertically.

ⓐ $\dfrac{1}{2} = \boxed{} = \boxed{} = \boxed{} = \boxed{}$

ⓑ $\dfrac{1}{3} = \boxed{} = \boxed{}$

ⓒ $\dfrac{1}{4} = \boxed{}$

Want to find

④ Let's refer to the number line and find fractions that are equivalent to one another other than those in ③.

⑤ Let's discuss what you have learned about the size of fractions and try to summarize the ideas.

 Summary

① When the denominators of fractions are the same number, they become larger as the numerator increases.

② When the numerators of fractions are the same number, they become smaller as the denominator increases.

③ Some fractions have the same size even if their denominators and numerators are different.

Want to confirm

1 ▶ Which one is larger? Let's fill in the ☐ with an equality or inequality sign.

① $\dfrac{3}{7} \boxed{} \dfrac{5}{7}$　　② $\dfrac{3}{5} \boxed{} \dfrac{3}{8}$　　③ $\dfrac{1}{2} \boxed{} \dfrac{4}{8}$

Want to think   Addition of fractions with the same denominator

1

I made milk coffee with $\dfrac{3}{6}$ L of coffee and $\dfrac{4}{6}$ L of milk. How many liters of milk coffee did I make in total?

$$\dfrac{3}{6} + \dfrac{4}{6} = \boxed{}$$

$$= \boxed{} \dfrac{\boxed{}}{\boxed{}}$$

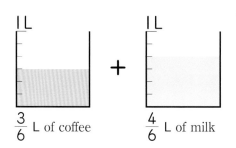

$\dfrac{3}{6}$ L of coffee $\dfrac{4}{6}$ L of milk

Way to see and think

Should we think how many sets of $\dfrac{1}{6}$?

When adding fractions with the same denominator, keep the denominator and add the numerators.

If we change this to a mixed fraction, the size is easier to understand.

Want to confirm

 Let's solve the following calculations.

① $\dfrac{2}{4} + \dfrac{1}{4}$ ② $\dfrac{4}{7} + \dfrac{1}{7}$ ③ $\dfrac{2}{8} + \dfrac{3}{8}$

④ $\dfrac{2}{3} + \dfrac{2}{3}$ ⑤ $\dfrac{2}{5} + \dfrac{4}{5}$ ⑥ $\dfrac{3}{9} + \dfrac{6}{9}$

Want to deepen

 In the calculation of $\dfrac{2}{4} + \dfrac{3}{4}$, Daiki made a mistake as shown on the right. What is wrong?

Let's write it correctly.

 Daiki's idea

Considering the following diagram,

$$\dfrac{2}{4} + \dfrac{3}{4} = \dfrac{5}{8}$$

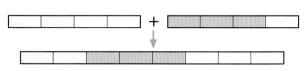

2

Let's think about how to calculate $1\dfrac{3}{5} + 2\dfrac{4}{5}$.

① Let's calculate by using the diagram on the right.

$$1\dfrac{3}{5} + 2\dfrac{4}{5} = 3\dfrac{7}{5}$$

$$= \boxed{}\dfrac{\boxed{}}{\boxed{}}$$

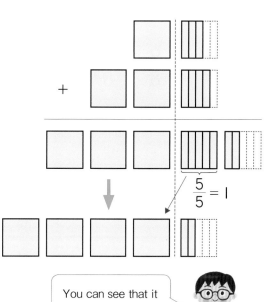

$$\dfrac{5}{5} = 1$$

② Let's calculate by changing to improper fractions.

You can see that it increased by 1.

Hiroto

When we add mixed fractions, add the whole number parts, add the fractional parts, and then combine the results. When the sum of the fractional parts becomes an improper fraction, carry the part of the whole number.
Also, we can add by changing mixed fractions to improper fractions.

3 Let's solve the following calculations.

① $1\dfrac{1}{3} + 2\dfrac{1}{3}$ ② $3\dfrac{2}{7} + 1\dfrac{3}{7}$ ③ $2\dfrac{2}{6} + 4\dfrac{1}{6}$

④ $3 + 3\dfrac{5}{6}$ ⑤ $1\dfrac{2}{3} + 2\dfrac{2}{3}$ ⑥ $2\dfrac{7}{9} + \dfrac{4}{9}$

⑦ $\dfrac{2}{7} + 4\dfrac{6}{7}$ ⑧ $2\dfrac{1}{5} + 3\dfrac{4}{5}$ ⑨ $\dfrac{1}{4} + 2\dfrac{3}{4}$

3

There are $\dfrac{9}{8}$ L of apple juice and $\dfrac{4}{8}$ L of orange juice.

How many more liters of apple juice are there than orange juice?

$$\dfrac{9}{8} - \dfrac{4}{8} = \boxed{}$$

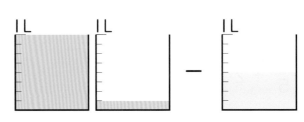

Way to see and think

How many sets of $\dfrac{1}{8}$ is the difference?

When subtracting fractions with the same denominator, keep the denominator and subtract the numerators.

 Let's think about how to calculate $3\dfrac{2}{3} - 1\dfrac{1}{3}$.

① Let's calculate by using the diagram on the

right.

$$3\dfrac{2}{3} - 1\dfrac{1}{3} = \boxed{}\dfrac{\boxed{}}{3}$$

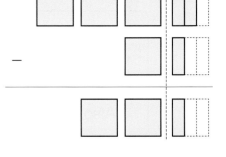

② Let's calculate by changing to improper

fractions.

When we subtract mixed fractions, subtract the whole number parts, subtract the fractional parts, and then combine the results.
Also, we can subtract by changing mixed fractions to improper fractions.

 Let's solve the following calculations.

① $\dfrac{5}{4} - \dfrac{3}{4}$ ② $\dfrac{10}{9} - \dfrac{8}{9}$ ③ $6\dfrac{5}{7} - 4\dfrac{3}{7}$

4

Let's think about how to calculate $3\dfrac{2}{5} - 1\dfrac{3}{5}$.

① Let's calculate by using the diagram on the right.

$$3\dfrac{2}{5} - 1\dfrac{3}{5} = 2\dfrac{\square}{5} - 1\dfrac{3}{5}$$

$$= 1\dfrac{\square}{5}$$

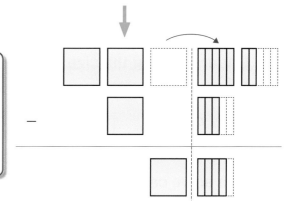

When the numerators of the fractional parts cannot be subtracted, calculate by borrowing 1 from the whole number part of the minuend.

② Let's calculate by changing to improper fractions.

6 Let's think about how to calculate $3 - 1\dfrac{1}{4}$.

7 Let's solve the following calculations.

① $1\dfrac{2}{4} - \dfrac{3}{4}$

② $1\dfrac{4}{9} - \dfrac{8}{9}$

③ $1\dfrac{1}{6} - \dfrac{2}{6}$

④ $6\dfrac{2}{7} - 4\dfrac{5}{7}$

⑤ $9\dfrac{3}{5} - 3\dfrac{4}{5}$

⑥ $7\dfrac{3}{8} - 4\dfrac{7}{8}$

⑦ $1 - \dfrac{1}{6}$

⑧ $8 - 1\dfrac{2}{7}$

⑨ $4 - 2\dfrac{1}{5}$

What you can do now

☐ **Understanding the relationship between mixed fractions and improper fractions.**

1 Let's answer the following problems.

① Let's represent the following length as a mixed fraction and an improper fraction.

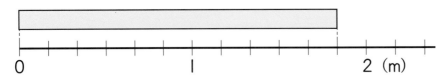

② Let's change mixed fractions to improper fractions and improper fractions to mixed fractions.

$$2\frac{3}{4} \qquad 3\frac{5}{6} \qquad \frac{7}{2} \qquad 4\frac{4}{9} \qquad \frac{7}{4} \qquad \frac{11}{5}$$

☐ **Understanding the size of fractions.**

2 Let's arrange the fractions inside the ( ) in descending order.

① $\left(\dfrac{1}{6}, \ \dfrac{1}{8}, \ \dfrac{1}{5}, \ \dfrac{1}{10} \right)$ ② $\left(2\dfrac{1}{8}, \ 2\dfrac{5}{8}, \ 2\dfrac{7}{8}, \ 2\dfrac{3}{8} \right)$

☐ **Can solve calculations with mixed fractions, improper fractions, and proper fractions.**

3 Let's solve the following calculations.

① $\dfrac{3}{4} + \dfrac{2}{4}$ ② $2\dfrac{1}{3} + 1\dfrac{1}{3}$ ③ $2\dfrac{2}{7} + 3\dfrac{5}{7}$

④ $1\dfrac{5}{8} + 1\dfrac{6}{8}$ ⑤ $\dfrac{11}{9} - \dfrac{4}{9}$ ⑥ $3\dfrac{5}{6} - 1\dfrac{4}{6}$

⑦ $5\dfrac{7}{15} - 3\dfrac{7}{15}$ ⑧ $4\dfrac{2}{7} - 1\dfrac{3}{7}$

☐ **Can make a math expression and find the answer.**

4 Yesterday at Takumi's house, they drank $1\dfrac{3}{5}$ L of milk in the morning and $\dfrac{4}{5}$ L of milk in the evening. Let's answer the following problems.

① How many liters did they drink altogether?

② Today, they drank $1\dfrac{2}{5}$ L of milk during the whole day. Between yesterday and today, in which day did they drink more milk? And by how many liters?

Supplementary Problems
•••••••• ➤ p.154

Usefulness and efficiency of learning

1 Which one is larger? Let's fill in each ☐ with an inequality sign.

☐ Understanding the relationship between mixed fractions and improper fractions.

① $3\frac{2}{3}$ ☐ $\frac{9}{3}$

② $\frac{10}{7}$ ☐ $1\frac{5}{7}$

③ $\frac{17}{6}$ ☐ $3\frac{1}{6}$

④ $3\frac{1}{4}$ ☐ $\frac{14}{4}$

☐ Understanding the size of fractions.

2 Let's solve the following calculations.

☐ Can solve addition and subtraction of fractions.

① $1\frac{2}{7} + 2\frac{2}{7}$

② $\frac{3}{5} + \frac{2}{5}$

③ $4\frac{2}{3} + 2\frac{2}{3}$

④ $2\frac{5}{9} + \frac{8}{9}$

⑤ $3\frac{4}{8} - 1\frac{3}{8}$

⑥ $1\frac{5}{9} - \frac{7}{9}$

⑦ $1 - \frac{7}{10}$

⑧ $4\frac{1}{5} - 2\frac{3}{5}$

⑨ $4\frac{1}{6} - 2\frac{5}{6}$

3 On Sunday, Hiroko ran $1\frac{2}{5}$ km in the morning and $1\frac{4}{5}$ km in the evening. How many kilometers did she run altogether? Also, how many kilometers is the difference?

☐ Can make a math expression and find the answer.

4 Let's look at the diagram shown on the right and answer the following problems.

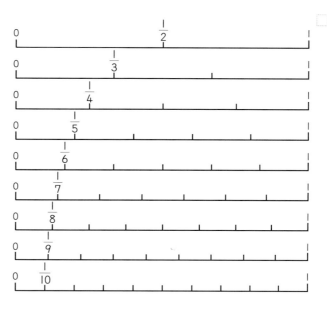

☐ Understanding the size of fractions.

① Let's write a fraction with the same size as $\frac{3}{5}$.

② Let's write the following fractions in descending order.

$\frac{5}{8}$, $\frac{5}{7}$, $\frac{7}{10}$

Box game

(There are squares)

Problem What are the properties of the shapes of the boxes?

18 Cuboid and Cube
Let's explore the properties of shapes of boxes and how to make them.

1 Cuboid and Cube

Want to classify Shape of boxes

1 Boxes were gathered from our surroundings. Let's classify them by investigating the shape of the faces of the boxes.

① Asahi classified the boxes as follows. How did she classify them?

Daiki

There are many rectangular faces.

Some are just square shapes.

Yui

101

A solid whose faces are only rectangles or rectangles and squares is called a **cuboid**.

A solid whose faces are only squares is called a **cube**.

A flat face like the faces of a cuboid and cube is called a **plane**.

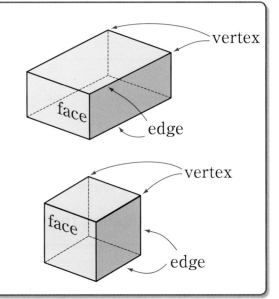

Want to explore

 Let's explore the faces, edges, and vertices of a cuboid or cube.

① Let's explore the number of faces, edges, and vertices, and summarize them in the table shown on the right.

	Cuboid	Cube
Number of faces		
Number of edges		
Number of vertices		

② In a cube, how many edges with an equal length are there?

③ How many edges are gathered at one vertex?

Want to explain

 Are the solids Ⓐ and Ⓑ shown on the right cuboids?

Also, let's explain the reasons.

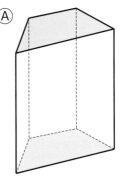

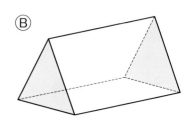

2 Nets

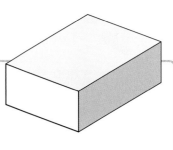

Want to know Cut open figure

1

What figure is formed when the cuboid box shown on the right is cut along the edges and opened?

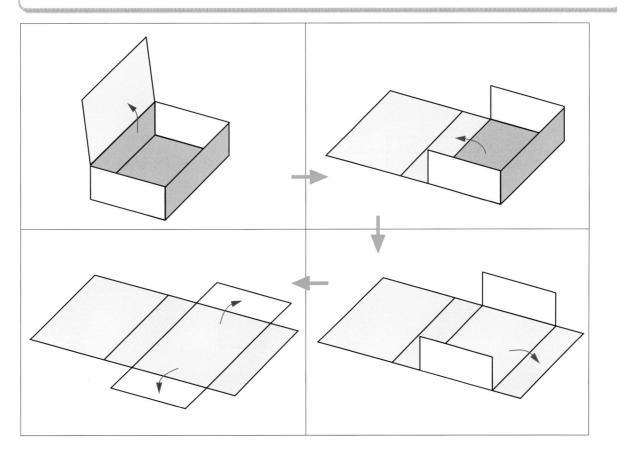

A figure drawn on a sheet of paper by cutting the edges of a box and unfolding it flat is called a **net**.

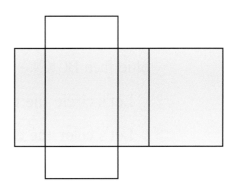

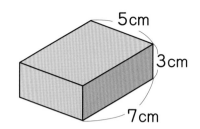

1 Let's think about what kind of net will be made when the cuboid shown on the right is made.

① We drew each of the six faces so that they could be assembled. Let's write down the length of ⓐ, ⓑ, and ⓒ.

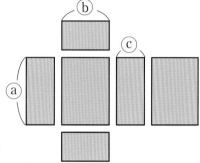

② Which of the following figures is the correct net? Let's confirm by using the faces drawn in ①.

Ⓐ   Ⓑ   Ⓒ

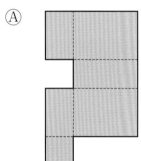

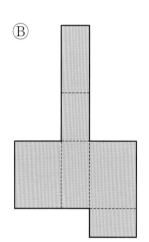

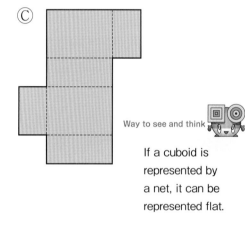

Way to see and think

If a cuboid is represented by a net, it can be represented flat.

2 We will make a cuboid by folding the net shown on the right. Let's answer the following questions about the cuboid.

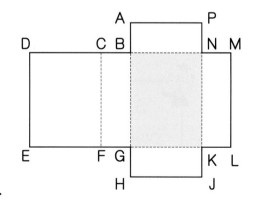

① Let's color the face opposite to the blue face BGKN.

② Let's circle the points that meet point M.

③ Let's color the edge that meets edge EF.

2

Let's make a cuboid as shown on the right.

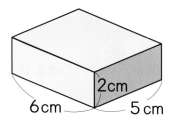

2cm

6cm   5cm

① Let's draw the rest of the net as shown below.

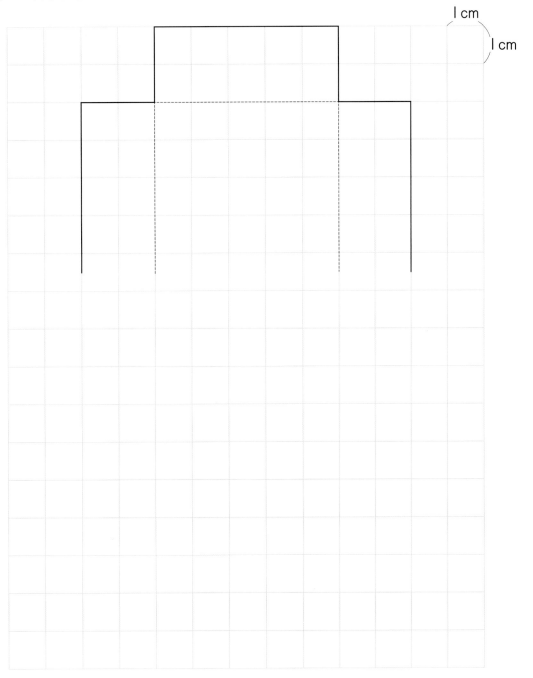

I cm

I cm

② Let's copy the net on cardboard, cut it out, and fold it.

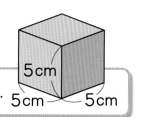

3

Let's think about the net of a cube with an edge of 5 cm.

① Nanami thought about net ⓐ and net ⓑ shown below.

Let's explain her idea.

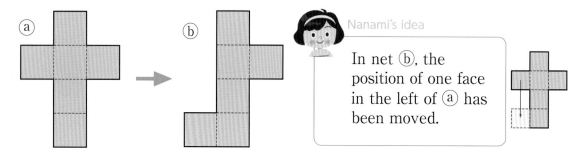

Nanami's idea

In net ⓑ, the position of one face in the left of ⓐ has been moved.

② Let's draw various nets based on the following nets.

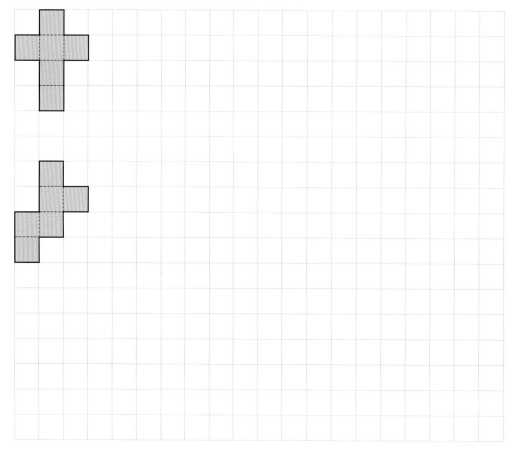

Are there 11 ways?

③ Let's present the nets made in ② to your friends.

3 Perpendicular and parallel on faces and edges

Want to know Relationship between faces

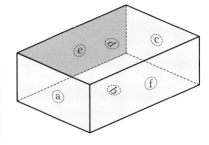

1 **Let's explore the relationship between the faces of the cuboid shown on the right.**

① Let's try to explore by placing a triangle ruler against the adjacent faces.

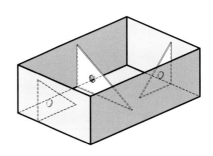

 Two adjacent faces of a cuboid are **perpendicular** such as ⓐ and ⓑ.

② Let's answer which faces are perpendicular to ⓒ.

③ Let's answer which face is not perpendicular to ⓐ.

 Two faces are **parallel** when they never intersect each other such as ⓐ and ⓒ, ⓑ and ⓓ, ⓔ and ⓕ in the above cuboid.

Want to confirm

 Let's answer the following questions about the cube shown on the right.

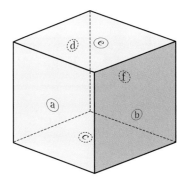

① Let's answer which faces are perpendicular to ⓐ.

② Let's answer which face is parallel to ⓔ.

Want to communicate

 2 Let's discuss why you can lay bricks neatly in any orientation, as shown on the right.

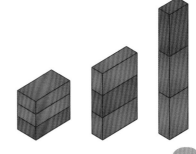

Relationship between edges

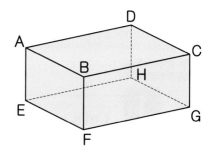

2 Let's explore the relationship between the edges of the cuboid shown on the right.

① How do edge AB and edge BF intersect? Let's check with a triangle ruler.

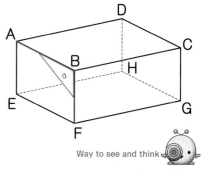

Way to see and think

What kind of figure is the face including edge AB?

Edge AB and edge BF are **perpendicular**.

② Let's find other perpendicular edges to edge AB.

③ What is the relationship between edge AB and edge EF?

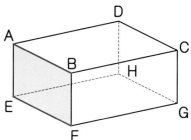

Edge AB and edge EF are **parallel**.

④ Let's find other parallel edges to edge AB.

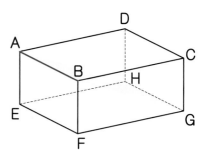

3 Which edges of the cube shown on the right are perpendicular to edge AB? Also, which are parallel edges?

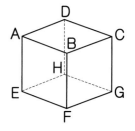

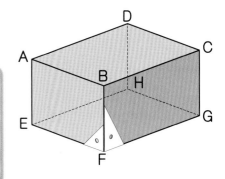

3 Let's explore the relationship between faces and edges of the cuboid shown on the right.

① Is edge BF perpendicular to face EFGH? Let's check with triangle rulers.

Edge BF and face EFGH are **perpendicular**.

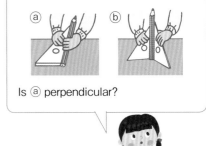

ⓐ   ⓑ

Is ⓐ perpendicular?

Yui

② Let's find other perpendicular edges to face EFGH.

③ Let's explain that edge AB is parallel to face EFGH.

Since face EFGH and face ABCD are parallel...

Daiki

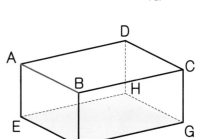

Edge AB and face EFGH are **parallel**.

④ Let's find other parallel edges to face EFGH.

4 In the classroom, let's look for the following faces and edges.

① Faces or edges perpendicular to the floor.

② Faces or edges parallel to the floor.

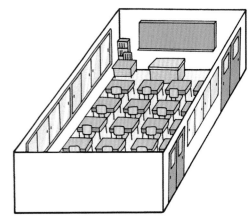

Want to know

1
Which of the following diagrams shows at a glance how the whole cuboid looks like?

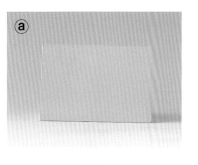

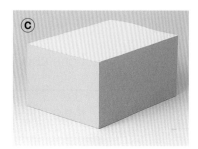

In ⓐ, I can see only a rectangle.

Daiki

In ⓒ, I can see many faces.

Nanami

Want to represent

 Let's draw a diagram so that you can see the whole cuboid at once.

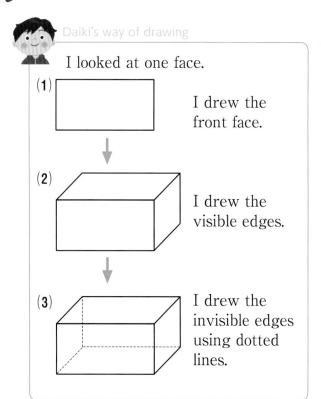

Daiki's way of drawing

I looked at one face.

(1) I drew the front face.

(2) I drew the visible edges.

(3) I drew the invisible edges using dotted lines.

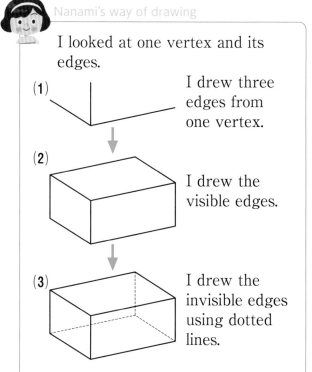

Nanami's way of drawing

I looked at one vertex and its edges.

(1) I drew three edges from one vertex.

(2) I drew the visible edges.

(3) I drew the invisible edges using dotted lines.

① Let's draw a cuboid by using either Daiki's or Nanami's way of drawing.

A diagram that is drawn to give a quick view of the whole object is called a **sketch**.

In a sketch, we draw parallel edges as parallel lines.

The size of a cuboid is represented by the **length**, **width**, and **height** that meet at the same vertex.

The size of a cube is represented by the length of an **edge**.

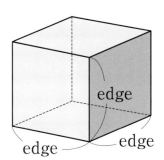

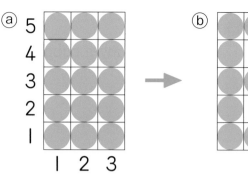

Want to know   How to represent positions on the plane

1

Let's think about the stones ● in ⓐ shown on the right.

① Let's remove two stones and design a symbol of **8**.

Can you explain to your friends which stones you removed?

Daiki

It's hard to communicate just with words.

Using the horizontal and vertical number...

Nanami

Purpose   How can we represent the position of things?

Want to explain

② Nanami represented the upper stone as (**2** and **4**) and the lower stone as (**2** and **2**). Let's explain the reasons.

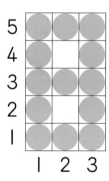

Summary

When expressing the position of an object on a plane, it can be represented by a pair of numbers, such as (2 and 4) and (2 and 2).

③ If you remove the stone at (**1** and **2**) on ⓑ, what symbol do the stones show?

④ Which stone on ⓑ should you remove to design the symbol **0**?

⑤ Let's make various symbols and tell your friends which stones you removed.

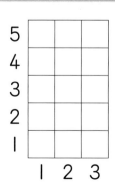

2 On the grid paper shown below, the vertical and horizontal axes are numbered. Point A is represented as (2 and 12).

Let's draw the following points in order and connect them with lines.

(2 and 12)→(6 and 15)→(10 and 15)→(13 and 17)→(13 and 15)→(18 and 11)→
(20 and 6)→(22 and 3)→ (19 and 4) → (16 and 3) → (18 and 6) →(15 and 9)→
(12 and 9)→(10 and 8)→ (11 and 6) → (10 and 6) → (7 and 8) →(4 and 6)→
(2 and 6) →(1 and 7) → (2 and 9) →(2 and 12)

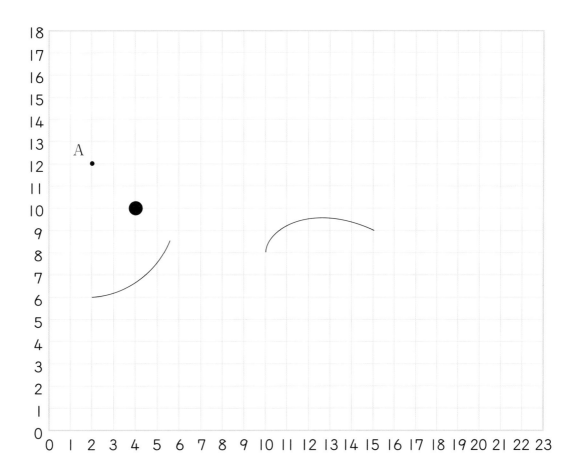

3   Based on the position of the flags, let's represent the position of animals by using numbers.

Every position in the space is represented by a list of three numbers.

For example, in the diagram below, the position of the monkey is 3 width, 1 length, and 2 height. We express the position as (3, 1, 2).

①　Let's represent the position of the following animals.

Fox（　　　　　　）

Rabbit（　　　　　　）

Chicken（　　　　　　）

②　What animal is in position (4, 1, 3)?

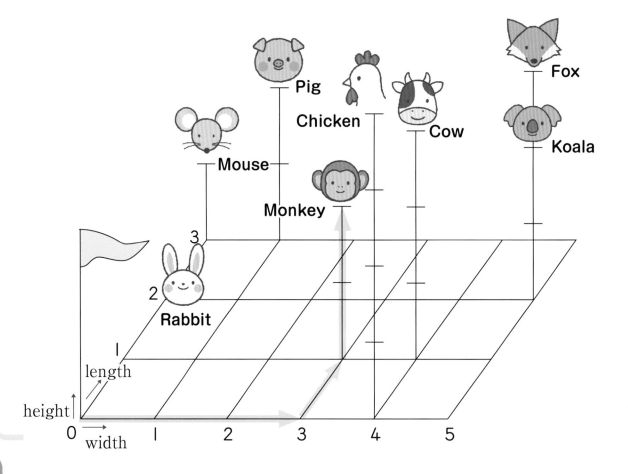

Let's search for things where the representation of positions is utilized.

① Braille

In some streets and buildings, you may find Braille. Braille uses a special ink or other so that they can read it by touching.

Braille consists of raised dots with two columns and three rows.

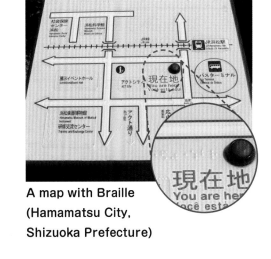

A map with Braille (Hamamatsu City, Shizuoka Prefecture)

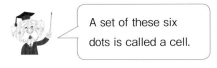
A set of these six dots is called a cell.

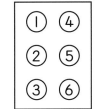

⑥ is in the third row from the top and second column from the left.

Daiki

あ (a)	い (i)	う (u)	え (e)	お (o)

② Shogi

In Shogi, the position of the piece is represented by a pair of numbers such as length and width or column and row. For example, the position of the piece moved in the diagram on the right is represented as "7六."

987654321

一二三四五六七八九

115

What you can do now

☐ **Understanding the properties of cuboids and cubes.**

1 Let's summarize about cuboids and cubes.

① Cuboids and cubes are classified by the shape of ☐ .

② Cuboids are only covered by ☐ or by rectangles and squares.

Cubes are only covered by ☐ .

③ The number of edges for both cuboids and cubes is ☐ .
The number of vertices for both cuboids and cubes is ☐ .

☐ **Can draw a net.**

2 Let's draw the net of a cube with an edge of 4 cm.

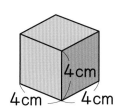

4 cm 4 cm 4 cm

☐ **Understanding perpendicular and parallel on faces and edges.**

3 The diagram on the right is a cuboid box. Let's answer the following problems.

① Which edges are perpendicular to edge AE?
② Which edges are parallel to edge AE?
③ Which face is parallel to face ABCD?
④ Which faces are perpendicular to face AEFB?

☐ **Understanding how to represent positions.**

4 On the diagram shown on the right, if the position of the fox is represented by (4 and 2), then how can you represent the position of the mouse and the rabbit?

Supplementary Problems
•••••••• ➤ p.155

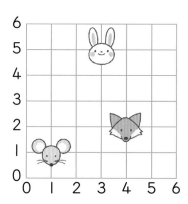

Usefulness and efficiency of learning

1 As shown below, there are a number of sheets of cardboard of different sizes. Make cuboids and cubes by using them. How many sheets of cardboard of each size are there in each solid?

☐ Understanding the properties of cuboids and cubes.

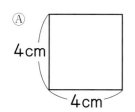

Ⓐ 4cm 4cm

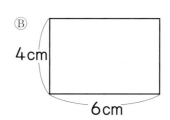

Ⓑ 4cm 6cm

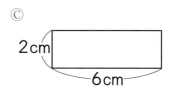

Ⓒ 2cm 6cm

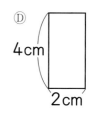

Ⓓ 4cm 2cm

2 As shown below, a net was designed to make a cube whose faces spell out the word "MATH." Let's write the missing letters in nets ① and ②.

☐ Can draw a net.

☐ Understanding perpendicular and parallel on faces and edges.

A T

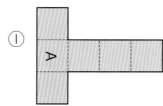

① A

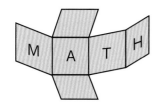

M A T H

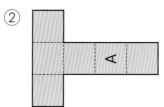

② A

3 As shown on the right, there is a cuboid. The positions of vertices A, E, F, G, and H are represented as follows.

A (1, 1, 4) E (1, 1, 0)
F (6, 1, 0) G (6, 4, 0)
H (1, 4, 0)

Let's represent the position of vertices B, C, and D.

☐ Understanding how to represent positions.

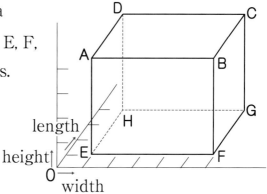

117

What is the relationship between two quantities?

Let's keep killifish in the classroom.

I have to wash the tank and fill it with water.

As time passes, the amount of water increases.

As time passes, the depth also increases.

There is likely to be a lot of quantities that change together.

I'm 10 years old and my brother is 7. When my brother's age increases, my age also increases. The difference is always the same, 3 years.

When the surrounding length is constant, if the length of the rectangle increases, then the width decreases.

None of them change in the same way.

Problem Are there rules for quantities changing together?

118

19 Quantities Changing Together
Let's investigate how two quantities change and their relationship.

Want to explore Relationship between length and width

1 Let's explore the relationship between length and width by drawing various rectangles and squares with a surrounding length of 16 cm.

Want to discuss

① Let's represent the length and width in the following table and discuss what you noticed.

Way to see and think
What is the rule between length and width?

Length and width

Length (cm)	1	2	3	4	5	6	7
Width (cm)							

Daiki

If the length increases, the width...

What kind of rules are there?

Yui

 Purpose Can we find the rules of two quantities changing together?

119

② How does the width change when the length increases by 1 cm?

You can see how the width correspondingly changes with the length.

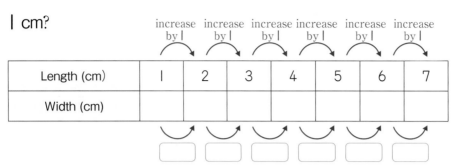

Length (cm)	1	2	3	4	5	6	7
Width (cm)							

③ Let's compare the sum of the length and the width.

④ Let's represent the rule between the length and the width in a math sentence. Consider the length as ☐ cm and the width as ◯ cm.

You're looking at the relationship between length and width.

Length + Width = 8

☐ + ◯ = 8

Summary

The rules of quantities changing together can be easily understood by representing the relationship in a table or math sentence.

 1

I read 10 pages of a 380-page book every day.

Let's answer the following questions.

① Let's represent the relationship between the number of read pages and the number of remaining pages in the following table.

Number of read pages and remaining pages

Number of read pages (pages)	10	20	30	40	50	60	70
Number of remaining pages (pages)							

② Consider the number of read pages as ☐ pages and the number of remaining pages as ◯ pages. Let's represent the relationship between ☐ and ◯ in a math sentence.

Mai's classroom is on the third floor.

By using the stairs, let's think about how to find the

height from the first floor to the third floor.

① As the number of steps increases, how does the height from the first floor change?

② There are 40 steps from the first floor to the third floor.

The height of each step is 15 cm. Let's represent the relationship between the number of steps and the height from the first floor in the following table.

Number of steps and height from the first floor

Number of steps (steps)	1	2	3	4	5	6	7
Height from the first floor (cm)	15	30					

There is not enough space to make the table for 40 steps.

Nanami

Can we find the rule between quantities in an easier way?

Hiroto

③ Let's find the rule between the number of steps and the height from the first floor. Consider the number of steps as □ and the height as ○ cm. Let's represent the relationship between □ and ○ in a math sentence.

Height of each step × Number of steps = Height from the first floor

☐ × □ = ○

④ Let's find the height for 40 steps by using the math sentence in ③.

3 We made the following figures by arranging some pieces of square paper with a side of 1 cm. Let's examine the relationship between the number of steps and the surrounding length of the figures.

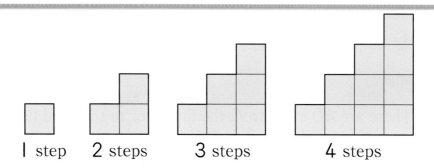

1 step   2 steps   3 steps   4 steps

①   How many cm is the surrounding length for 1 step, 2 steps, and so on? Let's represent it in the following table.

Number of steps and surrounding length

Number of steps (steps)	1	2	3	4	5	6	7
Surrounding length (cm)	4	8					

②   What rule can be found from the table in ①?

③   When the number of steps were considered as □ and the surrounding length as ○ cm, the relationship between □ and ○ was represented in the following math sentence. Let's explain how this was considered.

$$4 \times \square = \bigcirc$$

④   When the number of steps is 15, how many centimeters is the surrounding length?

⑤   When the surrounding length is 40 cm, what is the number of steps?

   Let's write two quantities changing together, □ and ○, that become into the following math sentence.

$$20 \times \square = \bigcirc$$

4 The following table shows how the pouring time and amount of water change as a bathtub is filled. Let's explore the relationship between the pouring time and the amount of water.

Pouring time and amount of water in the bathtub

Time (minutes)	0	2	4	6	8	10	12	14
Amount of water (L)	0	3	6	9	12	15	18	21

① Based on the table, let's draw points that represent the time and amount of water in the graph on the right.

② Let's connect the points with straight lines.

③ What is the amount of water, in liters, seven minutes after starting to fill the bathtub?

④ How many liters of water are there after 20 minutes?

⑤ Based on the table and graph, let's try to discuss what kind of rule there is on the relationship between time and amount of water.

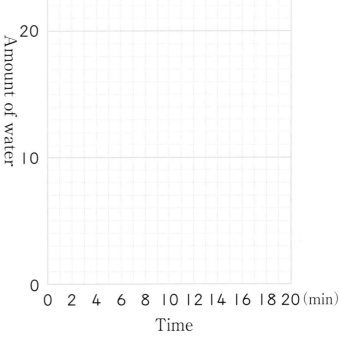

Pouring time and amount of water in the bathtub

If we make a graph, it's easier to understand how it changes.

What you can do now

☐ **Can represent the relationship between two quantities changing together in a table.**

1 Let's represent the relationship between the following two quantities in the tables below.

① The relationship between the length and width of a rectangle or square with a surrounding length of **36** cm.

Length and width

Length (cm)	1	2	3	4	5	6	7	8
Width (cm)								

② The relationship between the number of days and the remaining sheets of paper when **12** sheets of paper are used each day from a total of **96** sheets of paper.

Number of days and remaining sheets of paper

Number of days (days)	1	2	3	4	5	6	7	8
Remaining sheets of paper (sheets)								

☐ **Can represent the relationship between quantities changing together in a math sentence.**

2 There is a stone step with a height of **24** cm. Let's answer the following problems.

① Consider the number of steps as ☐ and the height from the ground as ○ cm.
Let's represent the relationship between ☐ and ○ in a math sentence.

② Let's find the height of the steps from the ground when there are **17** steps.

③ How many steps did you climb to reach the height of **18** m?

Supplementary Problems ▸ p.157

Usefulness and efficiency of learning

1 We connected 10 cm tapes as shown below. The length of each overlapping part is 1 cm. Let's answer the following problems.

☐ Can represent the relationship between two quantities changing together in a table.

① If we connect 2 pieces of the tape this way, how many centimeters is the total length?

② Let's write the numbers that apply in the table below.

Number of pieces of tape and total length

Number of pieces	1	2	3	4	5	6	7	8	9	
Total length (cm)	10									

③ If we connect 10 pieces of the tape, how many centimeters is the total length?

2 As shown below, trees are planted every 3 m. Let's answer the following problems.

☐ Can represent the relationship between quantities changing together in a math sentence.

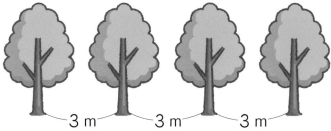

① When you plant five trees, how many meters is the distance from the first planted tree to the fifth one?

② Nanami represented the relationship between ☐ and ◯ in the following math sentence. She considered the number of planted trees as ☐ and the distance from the first planted tree as ◯ m. Let's explain what she thought.

$$3 \times (\square - 1) = \bigcirc$$

③ When planting 14 trees, how many meters is the distance from the first planted tree?

Let's deepen.

Can a relationship between any two numbers be represented in a math sentence?

Hiroto

Deepen.

Explore the height of several cups.

Every time a box is added to the stack, the height of the stack increases by the height of 1 box. In the case of cups, there are overlaps. Thus, we cannot find the total height of the cups by adding just the height of 1 cup.

Want to know

There are orange and blue cups with the same shape. The height of both cups is 8 cm. Let's investigate how the height changes when we stack these two kinds of cups.

Orange

Blue

Want to think

① From the bottom, 5 cups are stacked in the following order: orange, blue, orange, … What is the color of the cup at the top?

② As shown on the right, when **2** cups are stacked, the height is **10 cm**.

How many centimeters is the height when **5** cups are stacked?

Let's explore by representing the relationship in a table.

Number of cups and total height

Number of Cups	1	2	3	4	5
Total height (cm)					

③ After stacking several cups, the sum of the visible parts of the orange cups is **6 cm**. How many cups were stacked?

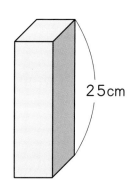

④ We want to place **5** blue cups and **5** orange cups in a box that has a height of **25 cm**.

Can we do that? Let's explain by using words and math expressions.

25cm

Daiki

If 2 cups are stacked, the height is 10 cm. If 3 cups are stacked, … We should think in order.

We already know the height for 5 cups …

Yui

How many times:  Length of the rubber

There are two rubbers, ⓐ and ⓑ.

Want to compare

1

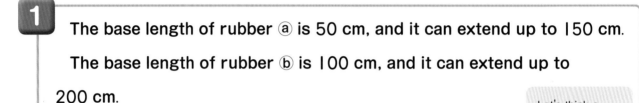

The base length of rubber ⓐ is 50 cm, and it can extend up to 150 cm.

The base length of rubber ⓑ is 100 cm, and it can extend up to

200 cm.

Let's compare how these two rubbers extend.

Let's think a comparison method for how the rubbers extend.

Nanami: Let's try to find the difference and compare.

Let's try to compare by "times of." — Hiroto

① Let's find the extension length of each rubber.

50cm

Rubber ⓐ ⟶ difference — 150cm

100cm

Rubber ⓑ ⟶ difference — 200cm

Rubber ⓐ $150 - 50 =$ ☐

Rubber ⓑ $200 - 100 =$ ☐

The extension length is the same.

The base length is different.

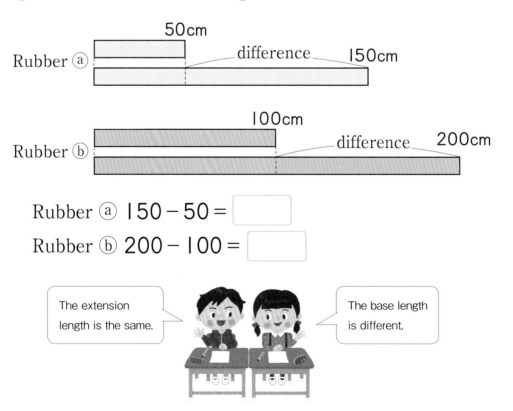

② Let's find how many times of the base length has been extended.

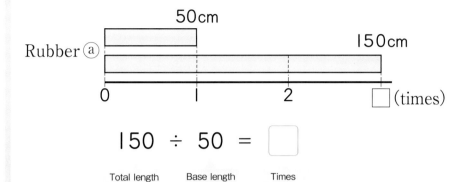

Base length | Total length

50 cm	150 cm
1 time	☐ times

Times

$$150 \div 50 = \boxed{}$$

Total length   Base length   Times

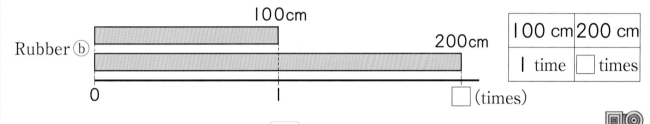

100 cm	200 cm
1 time	☐ times

$$200 \div 100 = \boxed{}$$

Total length   Base length   Times

③ From ① and ②, let's discuss which rubber extends better, ⓐ or ⓑ ?

Way to see and think

When you compare, the conclusion depends on whether you compare with the difference or "times of."

When a certain quantity is represented by how many times of the base quantity, "times of" is also called **ratio**.

When a weight was hung on spring Ⓐ, the length extended from 10 cm to 30 cm. When the same weight was hung on spring Ⓑ, the length extended from 5 cm to 20 cm.

Which spring extends better?

20 Utilization of Data
Let's interpret comparison graphs.

In mathematics class, we decided what we wanted to investigate in groups and we will present it on graphs, etc.

I ate ice cream yesterday, so I want to explore more about ice cream.

I found a foreign chocolate at a supermarket yesterday. How much chocolate does Japan buy from foreign countries?

Want to know

1 The following table summarizes how much money was used for ice cream by a family in Tokyo on every month in 2016. Let's think about this table.

Money used for ice cream by a family in Tokyo

Month	1	2	3	4	5	6	7	8	9	10	11	12
Amount of money (yen)	478	450	567	611	947	962	1309	1307	930	668	496	650

They bought a lot in July and August.

Some people bought ice cream even in winter.

Hiroto

Yui

① Let's represent the money used for ice cream in a bar graph.

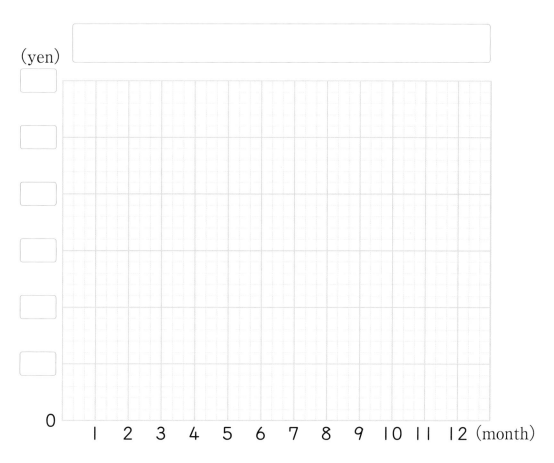

(yen)

0

1 2 3 4 5 6 7 8 9 10 11 12 (month)

② Let's look at the above bar graph and discuss what you noticed.

③ Looking at the above bar graph, Hiroto thought as follows.

Hiroto's idea

I think the money used for ice cream is related to the temperature.

Let's discuss a method to confirm Hiroto's idea.

④ In order to investigate the relationship between temperature and the money used for ice cream, we overlapped the graph of temperature in Tokyo in **2016** on the graph drawn in ①. Let's look at the graph below and discuss what you noticed.

A graph that overlaps two or more graphs in the same horizontal axis is called a composite graph.

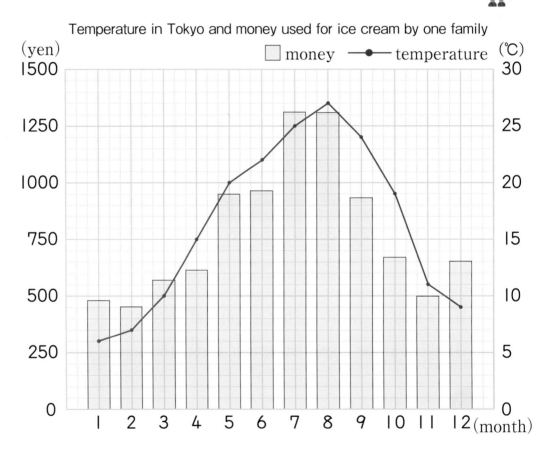

Temperature in Tokyo and money used for ice cream by one family

☐ money —●— temperature

When the temperature rises, you can see that the money used for ice cream is increasing. But, why is the money used for ice cream increasing in December?

Hiroto

The left vertical axis is amount of money and the right vertical axis is temperature.

2 The following table shows the amount of chocolate that Japan bought from foreign countries and the money paid for it. Let's think about this table.

Amount and money for chocolate that Japan bought from foreign countries

Can this be represented using both a bar graph and a line graph?

Daiki

Year	Amount (t)	Money (100 million yen)
2000	17959.69	154.03
2001	17330.82	148.40
2002	19091.23	161.21
2003	19550.17	161.35
2004	20322.91	167.55
2005	19887.45	173.42
2006	20312.27	188.68
2007	18864.50	188.40
2008	18550.54	176.73
2009	19374.84	162.08
2010	20801.77	164.62
2011	24013.94	178.79
2012	29751.09	217.47
2013	28996.13	266.26
2014	27657.13	290.96
2015	26323.96	293.34
2016	26141.07	272.85

① Let's look at the above table and discuss what you noticed.

That's a great amount.
Only with numbers, it's hard to know how they are increasing and decreasing.

Nanami

② Nanami represented the information from table in ① with the following composite graph. Let's look at this graph and discuss what you noticed.

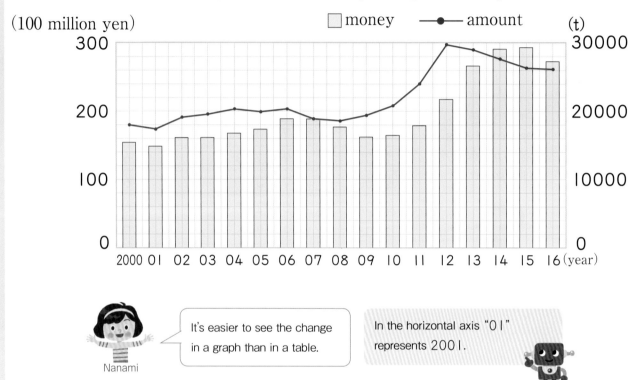

Amount and money for chocolate that Japan bought from foreign countries

It's easier to see the change in a graph than in a table.

Nanami

In the horizontal axis "01" represents 2001.

③ When Daiki investigated if Japan sells chocolate to foreign countries, he found a composite graph shown on the next page.

Let's look at this graph and discuss what you noticed.

Daiki

The amount sold in 2016 is about three times higher than in 2000. Japan also sells a lot of chocolate to foreign countries.

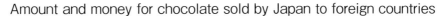

Amount and money for chocolate sold by Japan to foreign countries

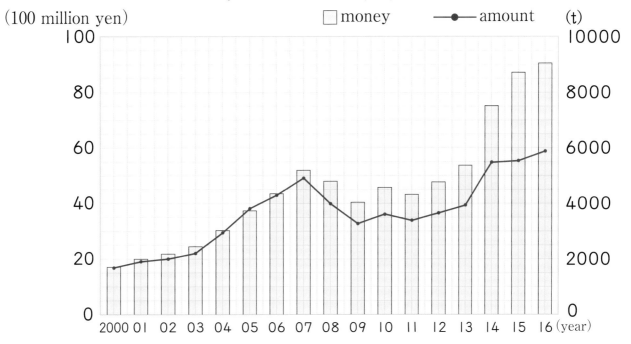

④ How many tons were sold to foreign countries in **2016**?

Also, about how many yen was that?

⑤ Let's compare graphs in ② and ③ to discuss what you can find.

The scale on the vertical axis is different.

It seems to be easier to understand if we align the scale on the vertical axis.

Japan bought more than four times the amount of chocolates sold to foreign countries in 2016.

The amount of money used in 2016 is about three times the money received.

21 4th Grade summary
Let's review learned mathematics.

Large numbers, round numbers

1 Let's read the following numbers. Also, let's round off to the nearest place written inside the [ ].

① 3824901 [ten thousands place] ② 64098172 [millions place]

③ 2715205860432 [ten billions place]

Large numbers, decimal numbers, fractions

2 Let's write the following numbers.

① The number that is the sum of 300 sets of one hundred million and 68 sets of ten thousand.

② The number that is the sum of 5 sets of 1, 3 sets of 0.1, and 2 sets of 0.001.

③ Mixed fraction and improper fraction that gathers 11 sets of $\frac{1}{7}$.

Decimal numbers, fractions

3 Let's write ↑ to point the following numbers in the number line below.

① 0.2   ② 1.6   ③ 2.35   ④ 3

⑤ $\frac{6}{10}$ ⑥ $2\frac{3}{10}$ ⑦ $1\frac{1}{10}$

0   1   2

Decimal numbers

4 Let's write the following numbers in descending order.

0.08   8   0.8   0.808   0

Decimal numbers, fractions, multiplication and division of decimal numbers

5 Let's solve the following calculations.

① $7.84 + 4.32$

② $6.89 + 5.3$

③ $8.4 - 2.01$

④ $\dfrac{3}{8} + \dfrac{7}{8}$

⑤ $2\dfrac{2}{7} + \dfrac{6}{7}$

⑥ $1\dfrac{7}{9} + 4\dfrac{7}{9}$

⑦ $1\dfrac{1}{3} - \dfrac{2}{3}$

⑧ $8\dfrac{1}{5} - 2\dfrac{3}{5}$

⑨ $3 - \dfrac{5}{6}$

⑩ 106×247

⑪ 0.61×8

⑫ 0.24×75

⑬ $96 \div 12$

⑭ $864 \div 36$

⑮ $1080 \div 72$

⑯ $75.2 \div 8$

⑰ $3.68 \div 16$

⑱ $45 \div 36$

Division by 1-digit number

6 There are 144 packages that will be placed in 3 trucks, each carrying the same number of packages. How many packages should one truck carry?

Division by 2-digit number

7 There are 127 children from 4th grade that will go to the top of a mountain using a cable car. Tha cable car can carry 25 people in one trip.

① How many trips will it take to carry all the children to the top?

② We want to carry as similar numbers of children as possible in each of 6 trips. How should the children be divided?

Math expressions and operations

8 The following calculations were solved. Let's find the mistake and change the right side of the equal sign.

① $10 - 3 \times 2 = 7 \times 2$
$= 14$

② $21 + 80 \times (13 - 7) = 101 \times 6$
$= 606$

9 How many degrees is the size of angle Ⓐ and Ⓑ?

10 Let's draw angles with size 70° and 123°.

11 Let's find the area of the colored parts.

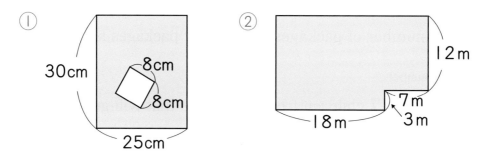

① 30cm 8cm 8cm 25cm

② 12m 7m 3m 18m

12 Let's write the number or unit that applies in the following ☐ .

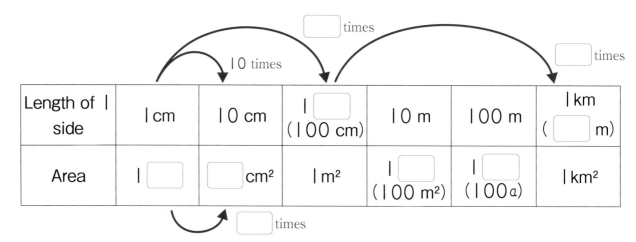

☐ times

10 times

☐ times

Length of 1 side	1 cm	10 cm	1 ☐ (100 cm)	10 m	100 m	1 km (☐ m)
Area	1 ☐	☐ cm²	1 m²	1 ☐ (100 m²)	1 ☐ (100 a)	1 km²

☐ times

13 How many degrees is the size of angles Ⓐ, Ⓑ, and Ⓒ shown on the right diagram?

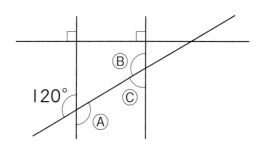

14 Let's draw the following quadrilaterals.

① Paralellogram

② Rhombus

③ Trapezoid

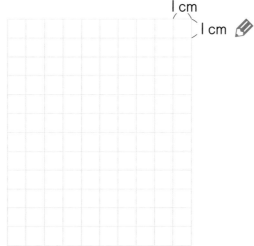

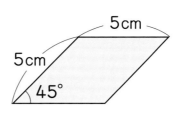

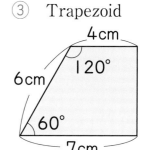

15 Let's make a cuboid box as the one shown on the right. Let's draw the nets of this box on grids Ⓐ and Ⓑ.

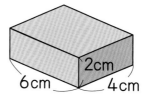

Ⓐ
1 cm
1 cm

Ⓑ
1 cm
1 cm

16 The following table shows the aluminum can production and the amount of cans that were recycled. Let's represent the data by round numbers in a line graph. What can you say from the graph?

Aluminum cans produced and recycled

Year	Aluminum cans produced (t)	Aluminum cans recycled (t)
2007	301451	279406
2008	299319	261338
2009	292897	273691
2010	296058	274242
2011	298224	275715
2012	301234	285401
2013	303830	254509
2014	312950	273491
2015	331500	298835
2016	341015	314965

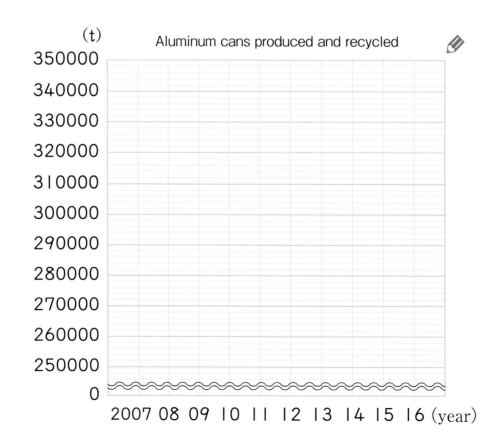

Aluminum cans produced and recycled

17 The line graph on the

right shows how the

temperature changes

in Tokyo and Sydney

in a year.

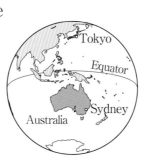

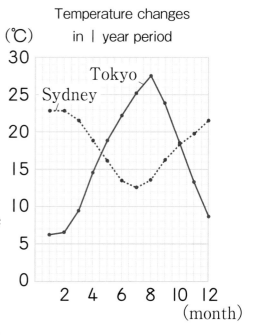

Temperature changes
in 1 year period

① Between which months is the temperature

in Tokyo higher than in Sydney?

② Which city has the highest change of

temperature?

Quantities changing together

18 There is a rectangle with a length of **4**cm. Let's explore how the

area changes as the width of the rectangle changes.

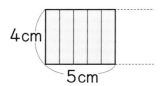

Width (cm)	1	2	3	4	5
Area of rectangle (cm²)	4	8			

① If the width of the rectangle increases by 1 cm , by how many cm²

does the area increase?

② When the area is **36** cm², how many cm is the width?

Secret of the calendar

On the calendar, choose a squared group of 9

numbers as shown on the right and calculate the

sum of those numbers. Did you find any secret? Try

another location. Do other locations on the calendar

have the same secret?

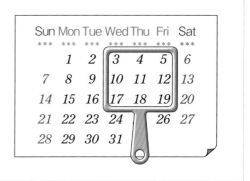

Computational thinking

Let's teach Robo "a method to draw a one-stroke sketch."

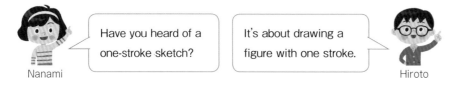

Nanami: Have you heard of a one-stroke sketch?

Hiroto: It's about drawing a figure with one stroke.

In one-stroke sketches you draw a line with the tip of a pencil without separating it from the paper. You may pass through the same point any number of times, but you must not trace the line that you once drew.

① Let's decide the starting point to make a one-stroke sketch of the figure shown on the right.

Nanami: I will try to start from point A.

Hiroto: I will try to start from point C.

② Let's teach Robo the process followed by the children in ① by using A, B, C, D, and E.

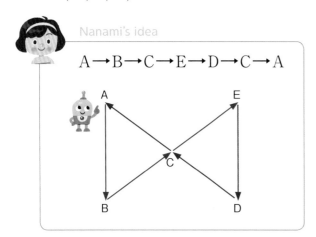

Nanami's idea

A→B→C→E→D→C→A

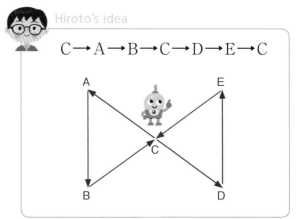

Hiroto's idea

C→A→B→C→D→E→C

③ Let's teach Robo a different drawing method from the ones already discussed by the children in **②**.

④ Let's teach Robo how to make a one-stroke sketch of the figure shown below.

①

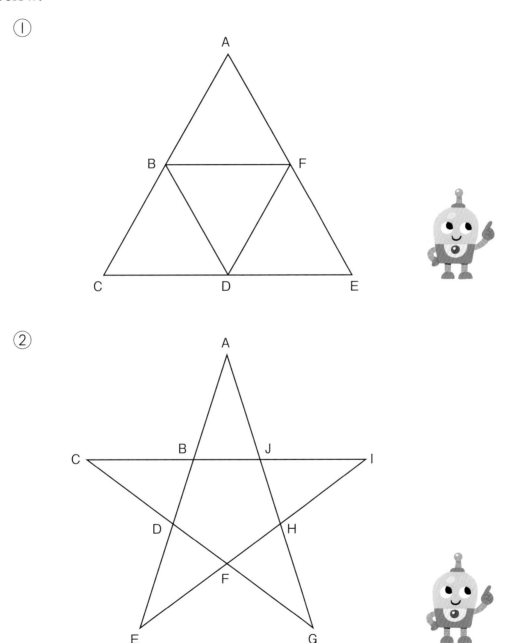

②

⑤ Let's find another figure that can be drawn through the one-stroke sketch and teach Robo how to draw it.

Utilize math for our life

Let's create an eco-friendly and kind school.

To create an eco-friendly and kind school, we thought about posters and plans.

Team that eliminates the waste of water

Hiroto

Our team investigated where and when water was being used.

< What we noticed >

· We looked at faucets. In two different places, the faucets were bad and water leaked.

· Drinking tap water from your mouth or hands will waste water. We think it's better to use a cup.

· In summer we have a swimming class at the school pool which use a lot of water. We think it's better to let people, from the region, use our pool if we use the same amount of water.

Amount of water used by one person every month

April	May	June	July	August	September	October	November	December	January	February	March
700 L	900 L	4500 L	5500 L	3500 L	2100 L	900 L	800 L	700 L	1000 L	1000 L	800 L

Let's represent this in a line graph and make a poster.

Team that separates trash

Yui

I think it's a good idea to place our own trash cans in the classroom. I thought about the design of the saparating trash boxes.

? Let's try to decide the size for the box designed by Yui and draw a corresponding net. Also, design your own trash can and try to draw its net.

Team that eliminates waste of school lunch

Daiki

At elementary schools and junior high schools nationwide, there seems to be about 17 kg of waste and leftovers for every person in a year. How much is 17 kg, considering various vegetables and other foods?

Food (weight in one serving)	Amount	Food (weight in one serving)	Amount
Eggplant (100 g)	170	Cucumber (100 g)	170
Potato (150 g)	113	Sausage (17 g)	1000
Egg (50 g)	340	Chicken thigh (250 g)	68

? Let's create a poster that replaces 17 kg with various vegetables.

Team for a kind school for blind people

Nanami

One way to tell blind people where to find things on top of a table is the clock position. It's a way of describing the position of things explained by the numerals in a clock face.

For example, in the figure on the left there is a compass at 8 o'clock.

Hiroto

? In the figure on the left, let's explain the position of each writing instrument using the clock position.

Utilize math for our life

Let's create an eco-friendly and kind school.

1. Toward learning competency.

	😊 Strongly agree	😐 Agree	🙁 Don't agree
① It was fun working on it.			
② The learning contents were helpful.			
③ Concentrated our efforts to make something better.			

2. Thinking, deciding and representing competency.

	😊 Definitely did	😐 So so	🙁 I didn't
① In making an eco-friendly and kind school, I was able to discover where to use the mathematical knowledge.			
② I was able to confirm if the idea could be realized by mathematics.			
③ I was able to represent my ideas in text, pictures, figures, and tables.			

3. What I know and can do.

	😊 Definitely did	😐 So so	🙁 I didn't
① I was able to make a better plan.			
② Making a plan deepened my understanding of mathematical knowledge.			

4. Encouragement for myself.

	😊 Strongly agree
① I think that I'm doing my best.	

Give yourself a compliment since you have worked so hard.

Let's try one more time, with a different theme, what you were not able to accomplish and keep doing your best on what you were able to fulfill.

Supplementary

Problems

10 Decimal Numbers ➡ p.148

11 Math Expressions and Operations ➡ p.149

13 Area ➡ p.150

15 Multiplication and Division of Decimal Numbers ➡ p.152

17 Fractions ➡ p.154

18 Cuboid and Cube ➡ p.155

19 Quantities Changing Together ➡ p.157

Answers ➡ p.158

⑩ Decimal Numbers

p.4~p.21

1 How many liters are the following amounts of water?

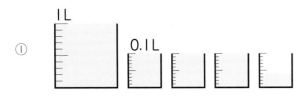

①

②

2 Let's represent the following measurements by using the unit shown in ( ).

① 2618 mm （m）

② 94 cm 5 mm （m）

③ 7328 m （km）

④ 15 km 462 m （km）

⑤ 4 kg 175 g （kg）

⑥ 637 g （kg）

3 Let's write the number that applies in the following ☐.

① $\frac{1}{10}$ of 0.1 is ☐.

② $\frac{1}{10}$ of 0.01 is ☐.

③ 0.01 is $\frac{1}{☐}$ of 1.

4 Let's write the number that applies in the following ☐.

① 5.248 is the number that is the sum of ☐ sets of 1, ☐ sets of 0.1, ☐ sets of 0.01, and ☐ sets of 0.001.

② The number that is the sum of 6 sets of 1, 9 sets of 0.1, and 1 set of 0.01 is ☐.

5 Let's write the following numbers.

① The number that gathers 45 sets of 0.01.

② The number that gathers 630 sets of 0.01.

③ The number that gathers 192 sets of 0.001.

④ The number that gathers 738 sets of 0.001.

6 Let's fill in the ☐ with the appropriate inequality signs.

① 0.376 ☐ 0.38

② 4.6 ☐ 4.571

③ 0.072 ☐ 0.207

7 Let's write the following numbers.

① The number that is 10 times of 0.58.

② The number that is 10 times of 1.334.

③ The number that is 10 times of 0.026.

④ The number that is $\frac{1}{10}$ of 7.69.

⑤ The number that is $\frac{1}{10}$ of 25.01.

8 Let's solve the following calculations.

① 3.12 + 5.34　②　3.29 + 4.36

③ 0.89 + 0.57　④　7.54 + 1.56

⑤ 8.06 + 1.84　⑥　4.37 + 2.83

9 Let's solve the following calculations.

① 6.42 + 3.3　②　4.9 + 2.34

③ 0.92 + 2.4　④　3 + 3.06

10 A ribbon was divided in two and resulted in two ribbons with lengths 1.43 m and 0.87 m. How many meters was the length of the original ribbon?

11 Let's solve the following calculations.

① 6.36 − 2.14　②　2.53 − 1.41

③ 5.74 − 2.27　④　7.09 − 2.35

⑤ 2.34 − 0.48　⑥　9.52 − 8.64

12 Let's solve the following calculations.

① 4.64 − 0.24　②　9.15 − 8.6

③ 3.07 − 1.98　④　3 − 1.74

13 There were 5.5 kg of rice, but 3.75 kg of it was eaten. How many kilograms of rice is remaining?

11 Math Expressions and Operations

p.22~p.35

1 Let's solve the following calculations.

① 500 − (320 + 70)

② 400 − (280 − 160)

2 I bought socks for 450 yen and a handkerchief for 400 yen, and paid with 1000 yen. How many yen was the change? Let's represent the problem with one math expression and find the answer.

3 Let's solve the following calculations.

① 15 + 4 × 3

② 56 − 36 ÷ 9

③ 10 ÷ 2 + 3 × 2

④ 28 ÷ 4 − 6 ÷ 2

⑤ 6 × 3 − 45 ÷ 5

4 Let's solve the following calculations.

① 45 ÷ 5 × 8

② 12 × 6 ÷ 4

③ (21 + 24) ÷ (35 − 26)

④ (8 + 3) × (6 − 2)

⑤ (98 − 18) ÷ 8 + 12

⑥ 86 − 49 ÷ (14 − 7)

5 Let's write the number that applies in the following ▢.

① $13 \times 4 + 17 \times 4$

  $= (13 + \boxed{}) \times 4$

  $= \boxed{} \times 4$

  $= \boxed{}$

② $29 \times 6 - 26 \times 6$

  $= (\boxed{} - 26) \times 6$

  $= \boxed{} \times 6$

  $= \boxed{}$

6 Let's solve the following calculations.

① $16 \times 7 + 14 \times 7$

② $33 \times 8 - 23 \times 8$

7 Let's write the number that applies in the following ▢.

① 199×9

  $= (\boxed{} - 1) \times 9$

  $= \boxed{} \times 9 - 1 \times 9$

  $= \boxed{} - 9$

  $= \boxed{}$

② 28×5

  $= (\boxed{} \times 2) \times 5$

  $= \boxed{} \times (2 \times 5)$

  $= \boxed{} \times \boxed{}$

  $= \boxed{}$

8 Let's solve the following calculations by using the rules of operations.

① 98×6

② 36×5

③ 101×27

13 Area

p.40~p.57

1 How many cm² is the area of the following figures?

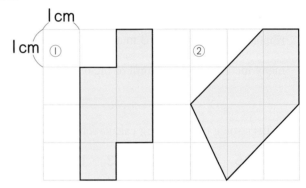

2 How many cm² is the area of the following rectangles?

①

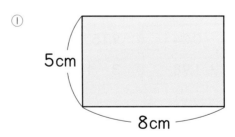

5cm   8cm

②

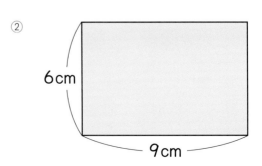

6cm   9cm

3 How many cm² is the area of the following rectangle and square?

①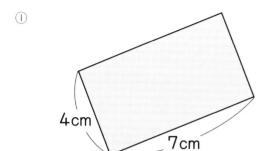

②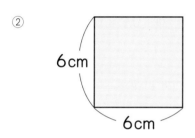

4 There is a rectangle with a width of 8 cm and an area of 48 cm². How many cm is the length of this rectangle?

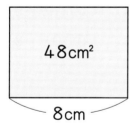

5 How many cm² is the area of the following figure?

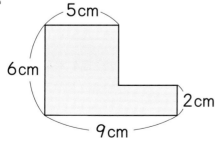

6 How many cm² is the area of the following figure?

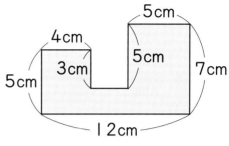

7 How many m² is the area of the following rectangle and square?

①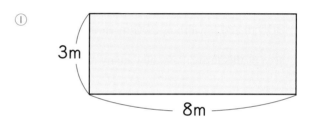

②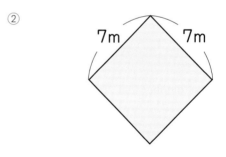

8 How many cm² is the area of the following rectangle?

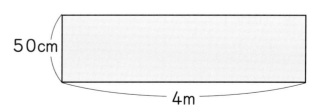

⑨ Let's write the number that applies in the following ☐.

① 800m² = ☐ a

② 4a = ☐ m²

③ 300a = ☐ ha

④ 9ha = ☐ a

⑤ 2ha = ☐ m²

⑥ 500000m² = ☐ ha

⑩ There is a rectangular farm with a length of 300 m and a width of 200 m. How many ha is the area of this farm? Also, how many a is it?

⑪ There is a square town with a side of 4 km. How many km² is the area of this town? Also, how many m² is it?

⑫ Let's write the number that applies in the following ☐.

① 7km² = ☐ m²

② 3000000m² = ☐ km²

③ 800ha = ☐ km²

④ 10km² = ☐ ha

⑮ Multiplication and Division of Decimal Numbers

p.65~p.77

▮ Let's solve the following calculations in vertical form.

① 2.4 × 2   ② 1.4 × 3

③ 1.6 × 4   ④ 3.6 × 2

⑤ 2.6 × 9   ⑥ 5.3 × 6

⑦ 0.8 × 4   ⑧ 0.9 × 6

② How many m² is the area of a rectangular flowerbed with a length of 3.7 m and a width of 5 m?

③ Let's solve the following calculations in vertical form.

① 3.6 × 5   ② 1.8 × 5

③ 2.5 × 6   ④ 0.5 × 8

④ Let's solve the following calculations in vertical form.

① 0.6 × 15   ② 1.2 × 31

③ 1.8 × 16   ④ 4.6 × 13

⑤ 6.7 × 17   ⑥ 8.6 × 19

⑦ 4.5 × 14   ⑧ 3.4 × 40

⑤ There are 12 bottles that contain 1.5 L of juice each. How many liters of juice are there in total?

6　Let's solve the following calculations in vertical form.

① 1.43×5　　② 0.89×6

③ 0.34×4　　④ 0.12×8

⑤ 0.06×5　　⑥ 0.35×4

7　A lake has a 1.75 km path around it. If you walk 4 laps around the lake, how many kilometers will you walk in total?

8　Let's solve the following calculations in vertical form.

① $6.4 \div 4$　　② $9.6 \div 8$

③ $8.4 \div 4$　　④ $6.9 \div 3$

⑤ $61.2 \div 18$　　⑥ $58.8 \div 42$

9　There is a rectangle with an area of 67.2 cm² and a length of 14 cm. How many centimeters is the width of this rectangle?

10　Let's solve the following calculations in vertical form.

① $2.4 \div 6$　　② $6.3 \div 7$

③ $1.38 \div 3$　　④ $3.75 \div 5$

⑤ $1.56 \div 4$　　⑥ $5.84 \div 8$

11　You will divide a 4.08 m tape into 6 tapes with equal length.

　　How many meters will the length of each tape be?

12　Let's solve the following calculations by dividing continuously.

① $6.4 \div 5$　　② $5.8 \div 4$

③ $9 \div 6$　　④ $6 \div 5$

⑤ $7 \div 4$　　⑥ $3 \div 8$

13　There is a square flowerbed with a surrounding length of 5.4 m. How many meters is the length of one side of this flowerbed?

14　Let's solve the following calculations. As for the quotient, let's find the nearest tenths place round number by rounding off the hundredths place.

① $4.7 \div 9$　　② $8.9 \div 7$

③ $5.6 \div 3$　　④ $9.4 \div 6$

⑤ $44.5 \div 13$　　⑥ $63.7 \div 97$

⑦ $75.6 \div 43$　　⑧ $34.5 \div 59$

15　If you divide a 8.5 m tape into 6 equal sections, about how many meters will each be? As for the quotient, let's find the nearest tenths place round number by rounding off the hundredths place.

16　16.5 L of juice will be poured into 3 L bottles. How many bottles of juice will be filled? How many liters of juice will remain?

⑰ Fractions

p.87~p.99

❶ How many meters is the length of the tape below? Let's represent it as an improper and a mixed fraction.

❷ Let's categorize the following fractions as proper, improper or mixed fractions.

Ⓐ $\dfrac{9}{8}$　　Ⓑ $\dfrac{3}{10}$　　Ⓒ $2\dfrac{3}{4}$

Ⓓ $\dfrac{6}{7}$　　Ⓔ $\dfrac{3}{3}$　　Ⓕ $1\dfrac{1}{2}$

❸ Let's write the following fractions represented by arrows A, B, and C as improper and mixed fractions.

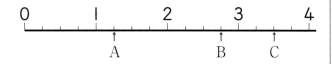

❹ Let's change the following mixed fractions to improper fractions.

① $4\dfrac{1}{2}$　　② $1\dfrac{8}{9}$

③ $3\dfrac{3}{4}$　　④ $2\dfrac{2}{7}$

⑤ $2\dfrac{2}{3}$　　⑥ $1\dfrac{7}{8}$

❺ Let's change the following improper fractions to mixed fractions or whole numbers.

① $\dfrac{9}{4}$　　② $\dfrac{11}{6}$

③ $\dfrac{10}{3}$　　④ $\dfrac{15}{5}$

⑤ $\dfrac{16}{7}$　　⑥ $\dfrac{16}{8}$

❻ Let's write the following fractions in ascending order.

① $\dfrac{7}{3}$, $1\dfrac{2}{3}$, $2\dfrac{2}{3}$

② $1\dfrac{7}{8}$, $\dfrac{13}{8}$, $\dfrac{19}{8}$, $2\dfrac{1}{8}$

③ $\dfrac{1}{5}$, $\dfrac{1}{7}$, $\dfrac{1}{3}$

④ $\dfrac{3}{5}$, $\dfrac{3}{4}$, $\dfrac{3}{8}$, $\dfrac{3}{2}$

❼ Let's solve the following calculations.

① $4\dfrac{3}{8} + 2\dfrac{4}{8}$　　② $3\dfrac{1}{5} + 5\dfrac{3}{5}$

③ $2 + 3\dfrac{3}{7}$　　④ $1\dfrac{5}{7} + 1\dfrac{3}{7}$

⑤ $3\dfrac{3}{5} + 1\dfrac{4}{5}$　　⑥ $3\dfrac{3}{4} + 2\dfrac{1}{4}$

⑦ $4\dfrac{7}{8} + \dfrac{4}{8}$　　⑧ $\dfrac{3}{4} + 3\dfrac{3}{4}$

⑨ $3\dfrac{2}{3} + \dfrac{1}{3}$　　⑩ $\dfrac{8}{9} + 1\dfrac{5}{9}$

⑧ Let's solve the following calculations.

① $\dfrac{8}{7} - \dfrac{2}{7}$

② $8\dfrac{2}{5} - 5\dfrac{1}{5}$

③ $7\dfrac{5}{9} - \dfrac{4}{9}$

④ $1\dfrac{1}{3} - \dfrac{2}{3}$

⑤ $1\dfrac{2}{7} - \dfrac{5}{7}$

⑥ $1\dfrac{2}{5} - \dfrac{4}{5}$

⑦ $3\dfrac{3}{8} - 1\dfrac{5}{8}$

⑧ $8\dfrac{1}{4} - 5\dfrac{3}{4}$

⑨ $1 - \dfrac{1}{5}$

⑩ $2 - \dfrac{2}{7}$

⑪ $6 - 1\dfrac{2}{5}$

⑫ $4 - 2\dfrac{2}{3}$

⑨ There are $2\dfrac{3}{8}$ L and $1\dfrac{5}{8}$ L of juice. Let's answer the following problems.

① How many liters of juice are there altogether?

② How many liters is the difference between the two amounts of juice?

⑩ The distance from Shu's house to school is $1\dfrac{3}{5}$ km, and the distance to the station is $2\dfrac{1}{5}$ km. How many kilometers is the difference between the two distances?

⑱ Cuboid and Cube

p.100～p.117

① Let's write the word that applies in the following ☐.

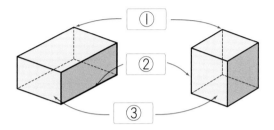

② Let's write the number that applies in the following ☐.

Cuboids and cubes have ☐① faces, ☐② edges, and ☐③ vertices.

③ Let's answer the problems below when the following net is folded to make a cuboid.

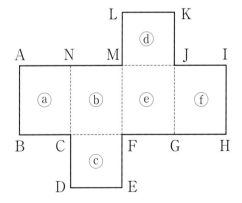

① Which face is opposite to face ⓐ?

② Which point overlaps with point A?

③ Which edge overlaps with edge DE?

4 From diagrams Ⓐ～Ⓕ, let's choose the nets that correctly fold into a cube.

Ⓐ
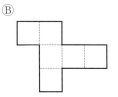

Ⓑ
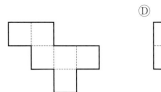

Ⓒ
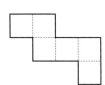

Ⓓ
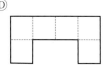

Ⓔ

Ⓕ

5 Let's answer the problems below when the following net is folded to make a cuboid.

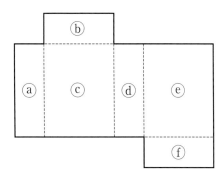

① Let's answer which faces are perpendicular to ⓐ.

② Let's answer which face is parallel to ⓑ.

6 Let's answer about the cuboid shown on the right.

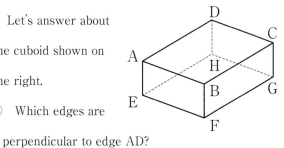

① Which edges are perpendicular to edge AD?

② Which edges are parallel to edge AE?

③ Which edges are perpendicular to face ABCD?

④ Which edges are parallel to face ABCD?

7 In the diagram below, the position of point A is represented by (4 and 8). Let's answer the following questions.

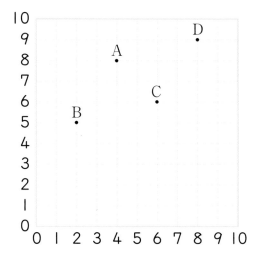

① Let's represent the position of points B, C, and D.

② In the above diagram, let's draw the following points: point E in (5 and 2) and point F in (9 and 4).

⑲ Quantities Changing Together

p.118~p.127

① Straws of the same length were used to create figures that aligns squares horizontally. Let's explore how the number of squares and the number of straws change.

1 square

2 squares

3 squares

① How many straws were used when figures with 1 square, 2 squares, and 3 squares were created?

② Let's write the numbers that apply in the following table.

Number of squares and number of straws

Number of squares	1	2	3	4	5	
Number of straws	4					

③ If the number of squares increases by 1, by how many will the number of straws increase?

④ How many straws were used when a figure with 8 squares was created?

② The following figures were made by arranging rectangles with a length of 1 cm and a width of 2 cm. Let's explore the relationship between the number of steps and the surrounding length.

Let's answer the following problems.

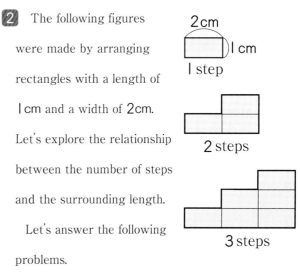

2 cm
1 cm
1 step

2 steps

3 steps

① For 1 step, how many centimeters is the surrounding length?

② For 2 steps, how many centimeters is the surrounding length?

③ For 3 steps, how many centimeters is the surrounding length?

④ How many times of the number of steps is the surrounding length?

⑤ Consider the number of steps as □ and the surrounding length as ○ cm. Let's represent the relationship between □ and ○ in a math sentence.

⑥ For 6 steps, how many centimeters is the surrounding length?

⑦ When the surrounding length is 54 cm, how many steps are there?

⑩ Decimal Numbers

❶ ① 1.34 L ② 0.47 L

❷ ① 2.618 m ② 0.945 m ③ 7.328 km
④ 15.462 km ⑤ 4.175 kg ⑥ 0.637 kg

❸ ① 0.01 ② 0.001 ③ 100

❹ ① 5, 2, 4, 8 ② 6.91

❺ ① 0.45 ② 6.3 ③ 0.192 ④ 0.738

❻ ① < ② > ③ <

❼ ① 5.8 ② 13.34 ③ 0.26 ④ 0.769 ⑤ 2.501

❽ ① 8.46 ② 7.65 ③ 1.46 ④ 9.1 ⑤ 9.9
⑥ 7.2

❾ ① 9.72 ② 7.24 ③ 3.32 ④ 6.06

❿ 1.43 + 0.87 = 2.3 2.3 m

⓫ ① 4.22 ② 1.12 ③ 3.47 ④ 4.74 ⑤ 1.86
⑥ 0.88

⓬ ① 4.4 ② 0.55 ③ 1.09 ④ 1.26

⓭ 5.5 − 3.75 = 1.75 1.75 kg

⑪ Math Expressions and Operations

❶ ① 110 ② 280

❷ 1000 − (450 + 400) = 150 150 yen

❸ ① 27 ② 52 ③ 11 ④ 4 ⑤ 9

❹ ① 72 ② 18 ③ 5 ④ 44 ⑤ 22 ⑥ 79

❺ ① 17, 30, 120 ② 29, 3, 18

❻ ① 210 ② 80

❼ ① 200, 200, 1800, 1791 ② 14, 14, 14, 10, 140

❽ ① 588 ② 180 ③ 2727

⑬ Area

❶ ① 6 cm² ② 7 cm²

❷ ① 40 cm² ② 54 cm²

❸ ① 28 cm² ② 36 cm²

❹ 6 cm

❺ 38 cm²

❻ 61 cm²

❼ ① 24 m² ② 49 m²

❽ 20000 cm²

❾ ① 8 ② 400 ③ 3 ④ 900 ⑤ 20000 ⑥ 50

❿ 6 ha, 600 a

⓫ 16 km², 16000000 m²

⓬ ① 7000000 ② 3 ③ 8 ④ 1000

⑮ Multiplication and Division of Decimal Numbers

❶ ① 4.8 ② 4.2 ③ 6.4 ④ 7.2 ⑤ 23.4
⑥ 31.8 ⑦ 3.2 ⑧ 5.4

❷ 3.7 × 5 = 18.5 18.5 m²

❸ ① 18 ② 9 ③ 15 ④ 4

❹ ① 9 ② 37.2 ③ 28.8 ④ 59.8 ⑤ 113.9
⑥ 163.4 ⑦ 63 ⑧ 136

❺ 1.5 × 12 = 18 18 L

❻ ① 7.15 ② 5.34 ③ 1.36 ④ 0.96 ⑤ 0.3
⑥ 1.4

❼ 1.75 × 4 = 7 7 km

❽ ① 1.6 ② 1.2 ③ 2.1 ④ 2.3 ⑤ 3.4 ⑥ 1.4

❾ 67.2 ÷ 14 = 4.8 4.8 cm

❿ ① 0.4 ② 0.9 ③ 0.46 ④ 0.75 ⑤ 0.39
⑥ 0.73

⓫ 4.08 ÷ 6 = 0.68 0.68 m

⓬ ① 1.28 ② 1.45 ③ 1.5 ④ 1.2 ⑤ 1.75
⑥ 0.375

⓭ 5.4 ÷ 4 = 1.35 1.35 m

⓮ ① 0.5 ② 1.3 ③ 1.9 ④ 1.6 ⑤ 3.4 ⑥ 0.7
⑦ 1.8 ⑧ 0.6

⓯ 8.5 ÷ 6 = 1.41… → 1.4 about 1.4 m

⓰ 16.5 ÷ 3 = 5 remainder 1.5 5 bottles, 1.5 L remain

⑰ Fractions

❶ $\frac{8}{5}$ m, $1\frac{3}{5}$ m

❷ Proper···Ⓑ, Ⓓ Improper···Ⓐ, Ⓔ Mixed···Ⓒ, Ⓕ

❸ A···$\frac{5}{4}$, $1\frac{1}{4}$ B···$\frac{11}{4}$, $2\frac{3}{4}$ C···$\frac{14}{4}$, $3\frac{2}{4}$

❹ ① $\frac{9}{2}$ ② $\frac{17}{9}$ ③ $\frac{15}{4}$ ④ $\frac{16}{7}$ ⑤ $\frac{8}{3}$ ⑥ $\frac{15}{8}$

❺ ① $2\frac{1}{4}$ ② $1\frac{5}{6}$ ③ $3\frac{1}{3}$ ④ 3 ⑤ $2\frac{2}{7}$ ⑥ 2

❻ ① $1\frac{2}{3}$, $\frac{7}{3}$, $2\frac{2}{3}$ ② $\frac{13}{8}$, $1\frac{7}{8}$, $2\frac{1}{8}$, $\frac{19}{8}$
③ $\frac{1}{7}$, $\frac{1}{5}$, $\frac{1}{3}$ ④ $\frac{3}{8}$, $\frac{3}{5}$, $\frac{3}{4}$, $\frac{3}{2}$

❼ ① $6\frac{7}{8}$ ② $8\frac{4}{5}$ ③ $5\frac{3}{7}$ ④ $3\frac{1}{7}$ ⑤ $5\frac{2}{5}$ ⑥ 6
⑦ $5\frac{3}{8}$ ⑧ $4\frac{2}{4}$ ⑨ 4 ⑩ $2\frac{4}{9}$

❽ ① $\frac{6}{7}$ ② $3\frac{1}{5}$ ③ $7\frac{1}{9}$ ④ $\frac{2}{3}$ ⑤ $\frac{4}{7}$ ⑥ $\frac{3}{5}$
⑦ $1\frac{6}{8}$ ⑧ $2\frac{2}{4}$ ⑨ $\frac{4}{5}$ ⑩ $1\frac{5}{7}$ ⑪ $4\frac{3}{5}$ ⑫ $1\frac{1}{3}$

❾ ① $2\frac{3}{8} + 1\frac{5}{8} = 4$ 4 L
② $2\frac{3}{8} - 1\frac{5}{8} = \frac{6}{8}$ $\frac{6}{8}$ L

❿ $2\frac{1}{5} - 1\frac{3}{5} = \frac{3}{5}$ $\frac{3}{5}$ km

18 Cuboid and Cube

1 ① Vertix ② Edge ③ Face

2 ① 6 ② 12 ③ 8

3 ① face ⓔ ② point I and point K ③ edge HG

4 Ⓐ, Ⓑ, Ⓒ, Ⓔ

5 ① ⓑ, ⓒ, ⓔ, ⓕ ② ⓕ

6 ① edge AB, edge AE, edge DC, edge DH

  ② edge BF, edge CG, edge DH

  ③ edge AE, edge BF, edge CG, edge DH

  ④ edge EF, edge FG, edge GH, edge HE

7 ① point B (2 and 5), point C (6 and 6), point D (8 and 9)

  ②

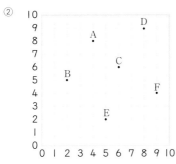

19 Quantities changing together

1 ① 1 square ⋯ 4 straws 2 squares ⋯ 7 straws

    3 squares ⋯ 10 straws

  ②

Number of squares and number of straws

Number of squares	1	2	3	4	5
Number of straws	4	7	10	13	16

  ③ 3 straws ④ 25 straws

2 ① 6 cm ② 12 cm ③ 18 cm ④ 6 times

  ⑤ $\square \times 6 = \bigcirc$ ⑥ 36 cm ⑦ 9 steps

Words and symbols from this book

Are	52		Net	103
Area	42		Parallel	107,108,109
Area of a rectangle	46		Perpendicular	107,108,109
Area of a square	47		Plane	102
Cube	102		Proper fraction	89
Cuboid	102		Sketch	111
Dividing continuously	72		Square centimeter	43
Formula	47		Square kilometer	54
Hectare	53		Square meter	50
Hundredths place	13		Tenths place	13
Improper fraction	89		Thousandths place	13
Mixed fraction	89			

1-cm² squares

▼ will be used in page 43.

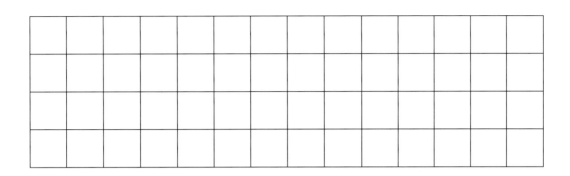

Blocks

▼ will be used in page 41.

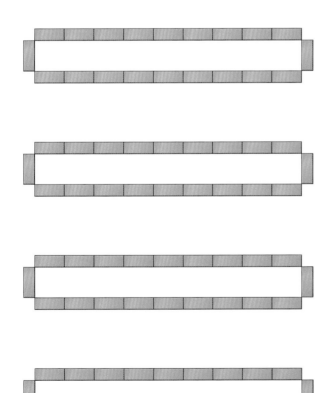

Memo

Editorial for English Edition:

Study with Your Friends, Mathematics for Elementary School
4th Grade, Vol.2, Gakko Tosho Co.,Ltd., Tokyo, Japan [2020]

Chief Editors:

Masami Isoda (University of Tsukuba, Japan), Aki Murata (University of Florida, USA)

Advisory Board:

Abrham Arcavi (Weizmann Institute of Science, Israel), Aida Istino Yap (University of the Philippines, Philippines), Alf Coles (University of Bristol, UK), Bundit Thipakorn (King Mongkut's University of Technology Thonburi, Thailand), Fou-Lai Lin (National Taiwan Normal University, Taiwan), Hee-Chan Lew (Korean National University of Education, Korea), Lambas (Ministry of Education and Culture, Indonesia), Luc Trouche (Ecole Normale Supérieure de Lyon, France), Maitree Inprasitha (Khon Kaen University, Thailand), Marcela Santillán (Universidad Pedagógica Nacional, Mexico), María Jesús Honorato Errázuriz (Ministry of Education, Chile), Raimundo Olfos Ayarza (Pontificia Universidad Católica de Valparaíso, Chile), Rogin Huang (Middle Tennessee State University, USA), Suhaidah Binti Tahir (SEAMEO RECSAM, Malaysia), Sumardyono (SEAMEO QITEP in Mathematics, Indonesia), Toh Tin Lam (National Institute of Education, Singapore), Toshikazu Ikeda (Yokohama National University, Japan), Wahyudi (SEAMEO Secretariat, Thailand), Yuriko Yamamoto Baldin (Universidade Federal de São Carlos, Brazil)

Editorial Board:

Abolfazl Rafiepour (Shahid Bahonar University of Kerman, Iran), Akio Matsuzaki (Saitama University, Japan), Cristina Esteley (Universidad Nacional de Córdoba, Argentina), Guillermo P. Bautista Jr. (University of the Philippines, Philippines), Ivan Vysotsky (Moscow Center for Teaching Excellence, Russia), Kim-Hong Teh (SEAMEO RECSAM, Malaysia), María Soledad Estrella (Pontificia Universidad Católica de Valparaiso, Chile), Narumon Changsri (Khon Kaen University, Thailand), Nguyen Chi Thanh (Vietnam National University, Vietnam), Onofre Jr Gregorio Inocencio (Foundation to Upgrade the Standard of Education, Philippines), Ornella Robutti (Università degli Studi di Torino, Iraly), Raewyn Eden (Massey University, New Zealand), Roberto Araya (Universidad de Chile, Chile), Soledad Asuncion Ulep (University of the Philippines, Philippines), Steven Tandale (Department of Education, Papua New Guinea), Tadayuki Kishimoto (Toyama University, Japan), Takeshi Miyakawa (Waseda University, Japan), Tenoch Cedillo (Universidad Pedagógica Nacional, Mexico), Ui Hock Cheah (IPG Pulau Pinang, Malaysia), Uldarico Victor Malaspina Jurado (Pontificia Universidad Católica del Perú, Peru), Wahid Yunianto (SEAMEO QITEP in Mathematics, Indonesia), Wanty Widjaja (Deakin University, Australia)

Translators:

Masami Isoda, Diego Solis Worsfold, Tsuyoshi Nomura (University of Tsukuba, Japan)
Hideo Watanabe (Musashino University, Japan)